BUILDING DRAINAGE

An Integrated Design Guide

Kemi Adeyeye and John Griggs

THE CROWOOD PRESS

First published in 2019 by
The Crowood Press Ltd
Ramsbury, Marlborough
Wiltshire SN8 2HR

www.crowood.com

British Library Cataloguing-in-Publication Data
A catalogue record for this book is available from the British Library.

ISBN 978 1 78500 639 5

Typeset by Sharon Dainton Design
Printed and bound in India by Parksons Graphics

Contents

Drainage systems are a small, but vital, part of the water cycle. Until recently, drainage design and engineering practices have been implemented in isolation of natural processes and cycles but this trend is beginning to change. The Integrated Water Management approach considers all waters to be a resource with value. So, whereas previously, water from appliances, homes, factories and roads was considered to be 'waste' that simply needed to be got rid of, it is now considered better to look at each element of the discharged water and consider how it might be utilized or reduced, or even eliminated.

In developed countries, the inhabitants have become disconnected with any processes beyond their homes. Toilets are used and the 'waste' is flushed away, out of sight and out of mind. Water comes from a tap that produces an endless supply. If light is wanted, a switch is thrown and to adjust the temperature in a building a dial is turned or a button pressed. This disconnection from the natural forces and resources has made people naturally less sensitive to environmental issues. However, the research and increased media attention to environmental issues has created a change of mind in many users, planners and engineers so that the impact on our world is now becoming part of any design process, from a simple house extension to the development of a new city.

It is against this backdrop of a more integrated approach to the use, reuse and management of water that this book is written. The authors have drawn on traditional engineering as well as the latest research to provide a useful design guide to drainage systems fit for the twenty-first century.

How to Use this Book

- This book is not intended to be read cover to cover before you can understand any particular topic. However, it can be read in sequence, as the book is arranged in a logical sequence based upon the conventional drainage chain.
- As you read the various sections, you will build knowledge and insight about drainage systems of the past, present and future. You will see how changing expectations, awareness and lifestyles have impacted on drainage design.
- The photos that are included do not always show good examples of drainage installations, but real drainage. Often drainage systems are designed well, but suffer from poor installation and lack of maintenance; often the illustrations reflect this as a sobering warning that drainage systems are not 'fit and forget' items.

- Within the text you will find various references to legislation, standards and other guidance documents. Do not expect these to be consistent; many will offer very different ways of designing a wastewater system based upon very different priorities. You will need to decide what the requirements and priorities of your system need to be to find appropriate guidance that reflects your needs.

- Many of the standards that are referenced are based on well-established fluid dynamics principles. This is a good starting point, but most wastewater composition is complex and so rarely behaves as a 'perfect fluid'. Although most drainage design equations will contain factors that aim to address the imperfection of wastewater and its pipework; these are often inadequate to deal with a system that ages and its operating conditions vary over time. Hence, most design principle are rules of thumb that need to be modified in the light to suit changing trends, as well as new and innovative products and fittings.

- The aim of the book is to make the reader think about any system, and not just follow a formula or a regulation.

- The issue of tolerances and accuracy is an important one that is touched upon in various places in the book. Some design parameters will be specified to tolerances of millimetres, others will have safety factors of two or three and some measurements will only be accurate within 200 per cent. So, designers are expected to understand and translate these guidelines to suit each individual project.

- Often you will find that different books, standards, regulations, etc. all use similar terms, but mean very different things. For example, a 'drain' in one country is defined as a ditch in another, and in another is a pipe of specific dimensions. Every effort has been made to ensure that this book is written in an accessible style so that the use of unexplained jargon is minimized and discussions are clear.

Fig. 0.1 The authors at work writing this book.

Water is a naturally occurring element on Earth: of which 97 per cent is salt water and is mainly unusable without expensive, large-scale desalination plants. The remaining percentage consists of about 69 per cent ice and snow cover, 30 per cent groundwater and less than 1 per cent surface water held in lakes and rivers (UNEP 2002).

The built environment contributes to climate change, resource availability and the resilience of human beings and their associated social and economic activities. It also contributes significantly to our physical and mental health and well-being in numerous ways.

Therefore, the efficient and effective design, construction, operation, maintenance and deconstruction of buildings and the built environment is both necessary and essential. Hence, it is so important that architectural professionals understand and holistically address sustainable practices, alongside other important design considerations such as space, form, materials, aesthetics, etc.

Modern sustainable buildings vary according to the level of resource use during construction and use in particular. Thus, the emerging characterizations and approaches to the production of buildings such as zero-carbon, low-carbon and carbon-negative buildings,

as well as autonomous buildings and Huf Hauss.

The main vehicle for the delivery of sustainable buildings is through building services. Building services consists of heating, cooling, lighting, ventilation and plumbing systems. Each of these can contribute to the overall sustainable performance of a building.

Therefore, it is essential for architectural designers to consider the following during the design of any building:

- The size of the building and the different activity areas within and around it.
- The internal spaces, zoning and layout
- Occupancy and use of the building
- Thermal and acoustic insulation
- The heating systems
- The cooling systems
- Ventilation – how fresh air is moved around the building
- Hot and cold water provision in the building
- Electricity supply, generation and possibly storage
- Lighting systems for the building
- On-site renewable systems
- Building management systems or controls.

It is noteworthy that architectural design, i.e. site, form, spatial and aesthetics requirements, etc. all impact on the

Left: Fig. 0.2 Water is central to our lives, for drinking, cooking, washing, heating, cooling as well as for aesthetics and recreation.

effectiveness of building services, and therefore the degree to which the building is sustainable and resource efficient.

This book is about water and wastewater drainage design in buildings. Water is used in the production, occupation and use of most buildings. Most building processes use water during on-site or off-site production, either through the production of dry walls in factories or through mixing mortar and concrete on site. Poor systems design can also result in supply issues, no or intermittent water coming out of the pipes, drips, water hammer and burst pipes due to excessive pressure, etc. In addition, site workers and building occupants use water directly for drinking, cooking, etc., and indirectly for cleaning, washing and waste disposal. The result of all of these activities is the production of waste or foul water. This by-product of water use needs to be appropriately 'handled' to avoid individual and public health and safety problems, e.g. foul odour, contamination and spread of diseases, etc., and to minimize the environmental impact such as the contamination of air, ground and water bodies.

The architectural and engineering design of the building needs to consider the type, positioning, access and management of drainage systems. In addition to delivering drainage systems that are both operationally effective and efficient, the decisions pertaining to water systems and drainage also impact on the degree to which the building is water efficient. This efficiency is increased by designing and specifying systems that help to reduce in-use water consumption, promote recycling, energy recovery and reduce the carbon/greenhouse, energy and environmental impacts associated with the collection, transportation, treatment and disposal of waste or foul water from buildings. Through good design, options for water conservation or reuse can be integrated into initial designs, product and fittings specification that would enable future reuse facilities to be incorporated simply at a later date. The routing of pipework can promote the waste of water (and energy) as users wait for hot water to arrive at the point of use. The pipe configuration can also enable or prohibit the immediate or future implementation of emerging innovations such as water recycling and reuse or heat recovery systems. The non-provision of storage, or its inadequate location or size within the building, basements or roof spaces during the design stages may also affect the ease of installation.

Early integration during the design stage is usually more beneficial, and cost effective, in building schemes or where facilities are shared between buildings, such as communal and district. Some building regulations now require integration with SuD (Sustainable Urban Drainage) systems, so architects need to understand the local situation and regulations to maximize the sustainability of buildings or a scheme.

In addition to building regulations, there are many environmental assessment schemes that will include water use in their assessment procedures. International schemes include BREEAM, LEED, Green Star, DGNB and Estidama. Many of these national and local schemes incentivize innovative design approaches and technologies. It may be a clause within local planning or other regulations that requires a new, or refurbished, building to achieve a certain level of a specified environmental assessment scheme. Often, one of the simplest ways of achieving assessment 'points' is through water efficiency measures and the drainage system will need to be designed to work with volumes of water that may be far lower than in a comparable building without water efficient measures.

This book addresses some of the important water and drainage design considerations and provides key knowledge and insights to aid effective design decision making. It is primarily targeted at architects, architectural technologists or architectural engineers but it would also be of interest to anyone who is curious about the hidden (and sometimes noisy) aspects of their homes and site, the benefits of having a good water and drainage systems and insights into what is happening when things go wrong.

The Water Cycle and Balance

The simple water cycle diagram in Fig. 0.3 that shows how water exists on the planet is familiar to most people. However, not everyone is fully aware of how human activity and interventions, in its various forms, affects the water cycle.

The amount of water used for human consumption in homes and places of work, in buildings, for agricultural, power generation, manufacturing and industrial processes, all impact on the built and natural environment. Too much can result in water ingress through roofs, flooding and landslides. Too little water in nature results in droughts, which affect water supply as well as drainage systems. Therefore, an understanding of the holistic water balance for all buildings, communities, rural or urban areas is paramount to make sure that we are resilient in this resource and avoid extreme human- or climate-in-

Water in	Consumption/Use	Wastewater out
Rainfall	Garden watering, vehicle washing, patio and path washing, ornamental ponds and water features	- Infiltration to local soil - Evaporation - Evapotranspiration via plants, or collection for reuse
Mains drinkable water	Drinking, culinary uses, showering, bathing	Blackwater and greywater to sewers or local treatment or collection for reuse
Mains non-drinkable water	WC flushing, hand washing	Blackwater and greywater to sewers or local treatment or collection for reuse
Recycled greywater	WC flushing	Blackwater to sewers or local treatment or collection for reuse
Captured rainwater	WC flushing, washing machines	Blackwater and greywater to sewers or local treatment or collection for reuse – as for rainfall
Captured condensate	WC flushing	Blackwater to sewers or local treatment or collection for reuse

Table 0.1 Detailed water balance.

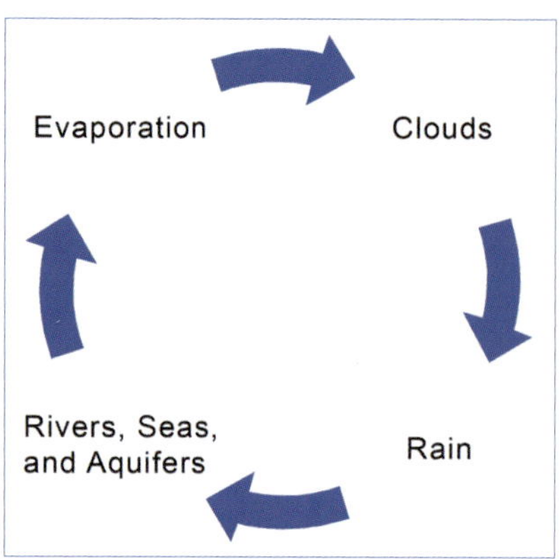

Fig. 0.3 The water cycle.

duced cycles of surpluses and deficits.

In its simplest form, the water balance is a measure of water in and wastewater out, but it can rapidly become more complex when contributions from rainwater or greywater, infiltration, or even embedded water, are considered. The balance between water in and wastewater out should be more or less equal. However, there is increasing imbalance of water in nature due to modernization, urbanization, population growth, and economic and industrial development. Technological innovation, however, makes it possible to address these deficits. For instance, it is now possible to recycle wastewater to drinking water standards. Thus we can become, in effect, water producers and not just consumers.

Table 0.1 shows that the options for 'wastewater out' are no longer simply to drain. This shows why drainage has become so important in the whole water cycle for domestic uses and many industrial processes. The options for reuse or recycling are becoming more viable and less expensive. This opens up possibilities to many users. British, European, American and International Standards now make provisions for various forms of recycling and reuse in buildings. Therefore, although the planet's overall use of water has increased, it is now possible, through good

design and implementation of systems, to minimize the impact of the ever-expanding built environment on the amount and quality of available water in our localities.

The production and discharge of wastewater raises a number of environmental issues. Firstly, the safe collection and transportation of the wastewater (through the sewer system), then the treatment of the wastewater to enable reuse, or safe discharge back into the natural system. The latter entails the use of significant amounts of energy and the associated carbon and greenhouse gas emissions. In addition to the energy consumption associated with the treatment of water and wastewater, there are increasing concerns about the amount of nitrates and sulphates (e.g. from soaps using in washing), medications, oils and fat and other non-biodegradable waste that end up in the wastewater system. In 2017, a report by CIWEM highlighted the issue of fibre loss from clothes. It explained that when clothes are washed they lose fibres, which are washed away into the drains and eventually make their way to the oceans, where they are consumed by fish and enter the food chain. This is particularly a problem with the volume of synthetic materials as well as plastics in the seas and the food chain. A quick examination of the filter on a tumble dryer will indicate how much fibre is lost during a typical wash. In fact, about 700,000 fibres can be released from a typical 6kg wash load. Fig. 0.6 from the report shows how wastewater contributes to the problem.

These complex chemicals, artificial fibres, medications and other contaminants are difficult to completely eliminate from wastewater. Therefore, residual amounts end up in water bodies and the natural environment. Studies have shown that these can potentially have an impact on marine and wildlife as well as the ecosystems that support them. Thus, it would be better if contaminants were removed before entering the drainage system. This can be achieved through public awareness, behaviour change and good design.

water in – water used/wasted = wastewater out

Fig. 0.4 Water balance equation.

Therefore, other 'man-made' cycles should be acknowledged in the natural water cycle. These should be taken into consideration in the design of any building drainage system. Fig. 0.7 shows some important design considerations depending on location and context.

Brief History of Drainage Systems

Since the beginning of time, rain from the sky has flowed overland and through the soil to rivers that lead to the sea. As civilizations developed, the rivers and nearby water bodies were utilized as disposal points for human waste and unwanted surface water, and sadly also for drinking water. As cities emerged, the open streams and ditches were covered and 'underground drains' created. Eventually, such drains were increasingly connected, enlarged and lined with brick or stone to produce a recognizable drain and sewer network.

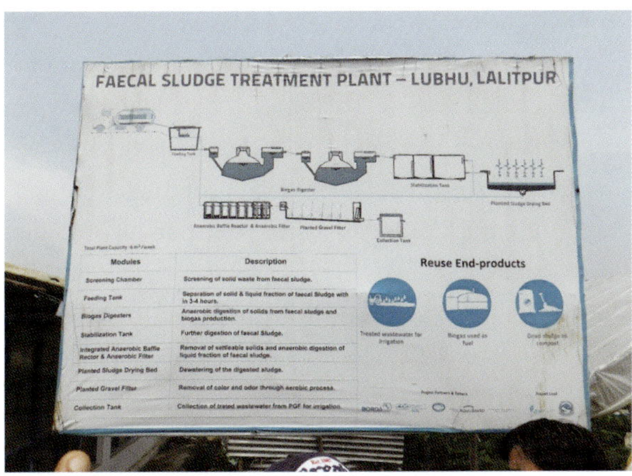

Fig. 0.5 Waste does not need to be discarded. In Nepal there is a plant that uses faecal sludge from the local wastewater treatment plants to produce irrigation water, compost and biogas fuel. Simple digesters, sand filters and settlement tanks are used on a hillside location to produce useful products with minimal additional energy.

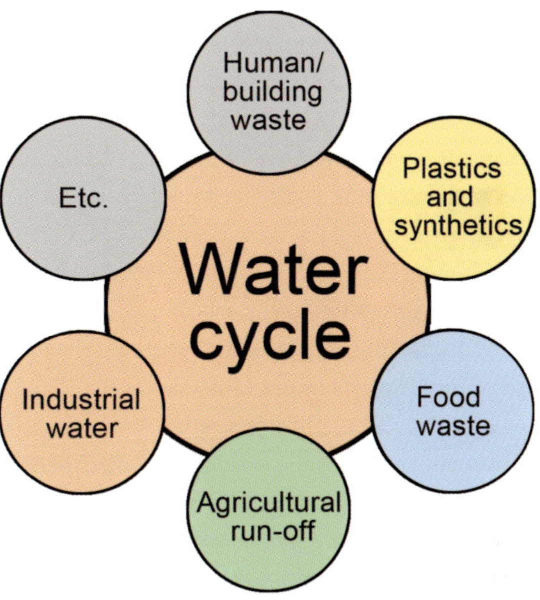

Fig. 0.7 Interconnected water cycles.

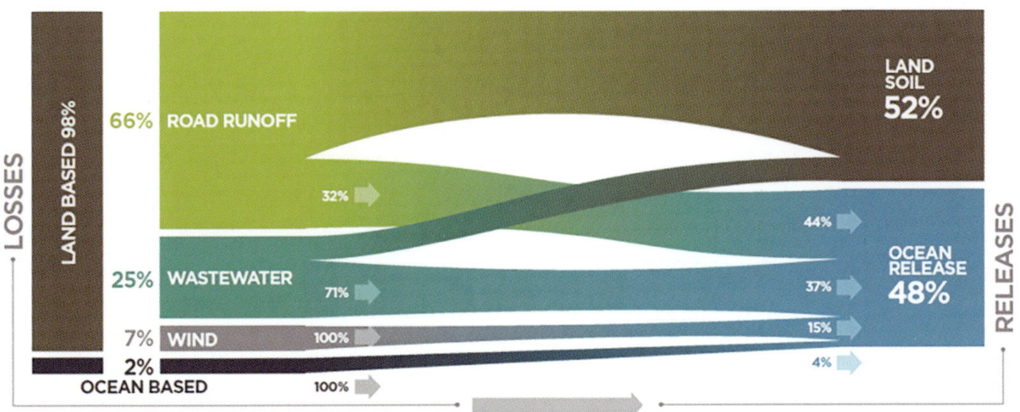

Fig. 0.6 Global releases to the world oceans. IUCN4. Contribution of different pathways to the release of microplastics. (Source: Boucher, J. and Friot D. (2017). Primary Microplastics in the Oceans: A Global Evaluation of Sources. Gland, Switzerland: IUCN. 43pp. Fig. 6.)

This process took place in various parts of the world at different times. One of the earliest sewerage systems has been found to date back to about 4000 BC in Babylon. It was used for removing stormwater from the cities. The Orkney Islands in the north of Scotland provides the example of an early foul water drainage system. Here, the total system only served about six dwellings, so it was not very extensive, but it was created around 3200 BC. At around the same time, some far larger systems were being developed in and around India.

One of the earliest recognized drainage systems was discovered in Asia in the Indus valley. Estimated to have been built around 2350 BC, the city of Lothal had an extensive drain and sewer system that included cesspits. Cities in the area were also built with sophisticated water supply systems and baths. It has been reported that they used rainwater harvesting systems and practised irrigation using wastewater.

Perhaps the most sophisticated wastewater systems of the time belonged to the Minoans on the island of Crete, which were used from around 3000 BC to 1000 BC. They utilized spigot and socket joints in earthenware pipes that were sealed using cement. They had flush toilets and practised rainwater harvesting.

The Romans were famous for their water supply and underfloor heating systems. Their sewer systems date back to around 800 BC. The purpose was mainly for surface water and marshland drainage, which evolved to remove human waste by 600 BC. However, houses were not connected to sewers in Rome until about 100 BC, so progress was rather slow.

Iron pipes were created in Germany during the fifteenth century to transport water and wastewater. John Harrington invented the flush toilet in England in the sixteenth century. Even though it was reported that the king had a sewer installed at the palace to take waste from the kitchen by the fourteenth century, it took until the seventeenth century before internal drainage systems were common in British castles and other buildings inhabited by the rich. However, in London, drain and sewer systems were evolving from the sixteenth to seventeenth centuries. Each area would develop its own system with no regard for any adjacent areas. King Henry VIII instigated a 'commission of sewers' to address this issue and gave people responsibilities for looking after and cleaning their local sewers. After the 'Great Stink' of 1858, a more radical solution to the problem of odour and pollution was sought urgently.

In the USA, it was not until 1728 that New York turned its open sewers into covered ones.

In the French capital, Paris, a series of cholera outbreaks in the 1830s resulted in the installation of large sewers between 1840 and 1890. These 'Les egouts' were so impressive that regular weekend boat tours were conducted. By the 1930s, the whole city had a combined sewer system installed.

This brings us to today. In many ways, drainage has not evolved or developed much over many centuries. However, with a growing population there is a need to ensure that not only is our drainage system efficient, but also safe. The European Commission released a press release that included this statement in 2018:

Water reuse in the EU today is far below its potential despite the fact that the environmental impact and the energy required to extract and transport freshwater is much higher. Moreover, a third of the EU's land suffers from water stress all year round and water scarcity remains an important concern for many EU Member States. Increasingly unpredictable weather patterns, including severe droughts, are also likely to have negative consequences on both the quantity and quality of freshwater resources. The new rules aim to ensure that we make the best use out of treated water from urban wastewater treatment plants, providing a reliable alternative water supply. By making non-potable (not for drinking) wastewater useful, the new rules will also contribute to saving the economic and environmental costs related to establishing new water supplies.

The rules it refers to are for irrigation of agricultural land using wastewater. The press release goes on to show how an integrated policy for water and wastewater is being developed based upon global needs and aims to:

Alleviate water scarcity across the EU, in the context of adapting to climate change. It will ensure that treated wastewater intended for agricultural irrigation is safe, protecting citizens and the environment. The proposal is part of the Commission's 2018 Work Programme, following up on the Circular Economy Action Plan, and completes the existing EU legal framework on water and foodstuffs. It complements the ongoing modernization of the European economy, the Common Agricultural Policy and climate change ambitions, and contributes to reaching the UN Sustainable Development Goals in the EU (in particular Goal 6 on water and sanitation), as well as fitting into the transition towards a Circular Economy which is an important goal of the Commission.

Thus, it appears that there are now policy drivers to ensure the integrated drainage design is considered with all other aspects of water to meet the needs of the planet and not just our local environment

The Drainage Chain

Water drainage can be defined as the safe and efficient removal of water and effluent from the point of origin to point of disposal without compromise to the health and safety of people in and around the building. Drainage systems help to deliver public health, minimize flooding and reduce contamination of the built environment.

Therefore, drainage in and around buildings serve three main purposes:

- To manage rainwater that falls on the roof
- To manage surface water on the ground around the site
- To manage foul/wastewater generated from inside the building.

Adequate site drainage is also needed to manage surface water; to avoid flooding as well as foul water encroachment. However, the process of doing this is not always straightforward. For example, there are currently about eighty-nine British Standards (BSs) relating to drainage! It will not be possible to cover all of this in this book. Therefore, the main regulations and standards will be signposted and the next chapter will cover the main design principles. But first, useful defi-

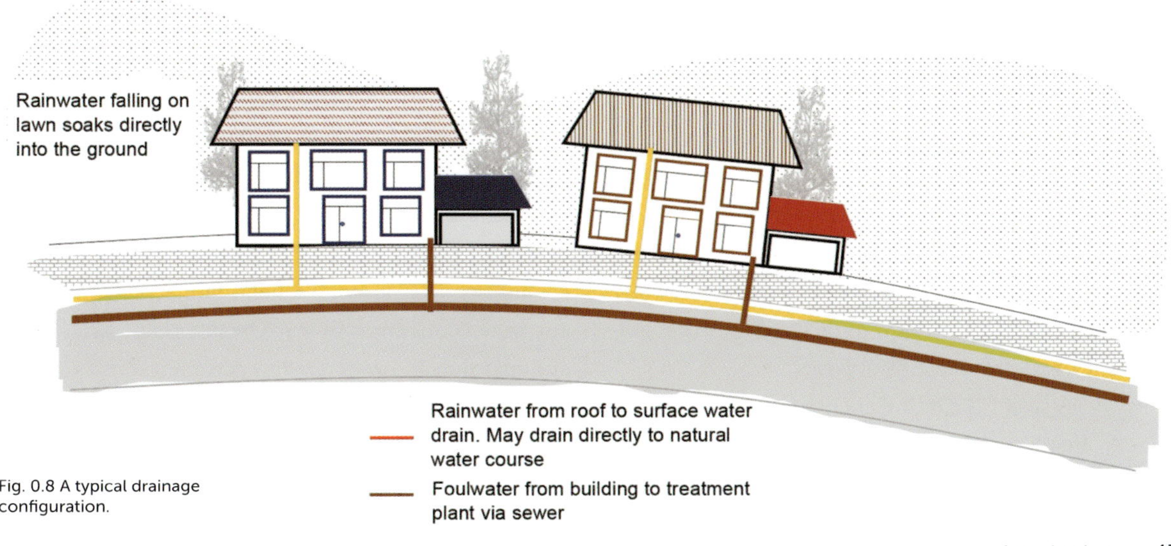

Fig. 0.8 A typical drainage configuration.

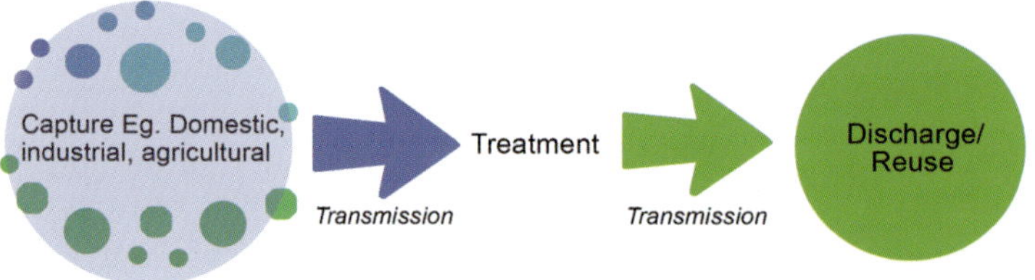

Fig. 0.9 Water drainage stages.

nitions:

- Domestic wastewater: water discharged from kitchens, laundry rooms, lavatories, bathrooms, toilets and similar facilities.
- *Drain: Typically, underground clay or plastic pipes that are used to carry wastewater from a source to a sewer network, and subsequently for treatment.
- *Drainage system: a connected natural or artificial network of pipes used to transport wastewater away from the source to final treatment or disposal sites.
- *Effluent or trade effluent: term used to describe wastewater from industrial activity.
- Foul water: wastewater from toilet, sinks, washing machines, dishwashers, etc..
- Lateral drain: sewer including ancillaries (access chambers, manhole, etc.) which serves a single property but lies outside that property's boundary, i.e. within or beneath another property boundary or the street.
- *Wastewater: water-composed by-product from domestic and non-domestic buildings, surface water run-offs and other intentional or non-intentional water and sewerage discharge into natural or artificial drainage systems.
- Wastewater treatment: is a natural, chemical and/or biological process to make wastewater and effluent safe for reuse or safe discharge to the environment.
- Water recycling and reuse: occurs when wastewater is captured and/or treated to be used for other domestic or non-domestic purposes.
- Sewer: A sewer including ancillaries (access chambers, manhole, etc.) is defined as a drain that is shared or used by more than one property.

- *Surface water: naturally occurring water from rainfall and precipitation, which has not seeped into the ground and which is discharged to the drain or sewer system directly from the ground or from exterior building surfaces.

*Definitions sourced from BS EN 1085:2007, superseded by BS EN 16323:2014 Glossary of wastewater engineering terms. Definitions may be paraphrased.

Drainage systems are used to manage surface and subsurface water, as well as effluent originating from human, industrial and agricultural activities. Contemporary drainage systems and processes broadly consist of three stages: Capture (source processes), treatment and discharge/reuse (destination processes) (Fig. 0.9).

Water Retention, Recycling and Reuse in the Drainage Chain

The introduction of water recycling and reuse into the drainage chain can make it a more sustainable process. The stages become circular, rather than linear, as shown in Fig. 0.10. Further, the journey for wastewater is significantly shortened as most of this is collected, treated (as necessary) and retained or reused closer to the point of origin.

Studies and trials have argued for and against the economies of scale, cost-benefit, return on investment, customer/consumer buy-in, health risks, ease of use and maintenance, energy and carbon intensity and so on, of such systems. What is obvious is that there can be economic, environmental and social benefits to a more circular approach to localized or decentralized

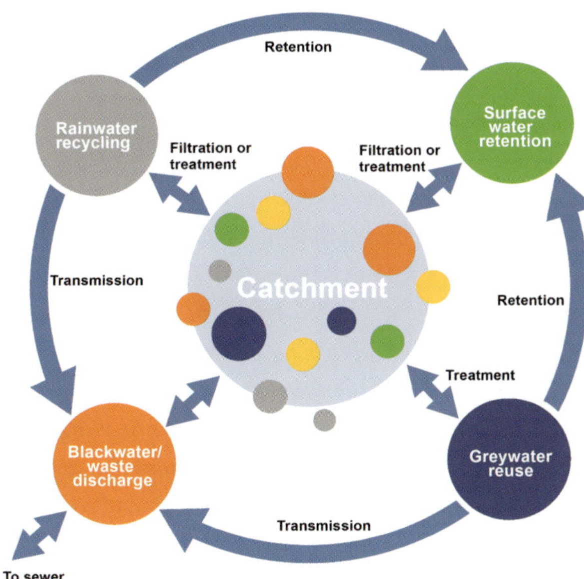

Fig. 0.10 Water drainage stages.

water systems in certain contexts or environment even if not applicable in all instances. Therefore, any worthwhile water and drainage design approach should consider water retention, recycling and reuse as part of a holistic water efficiency and resilience approach.

Alternative water supply and drainage systems will be discussed in more detail in subsequent chapters.

Drainage Standards and Regulations

An effective drainage system is important for public health as well as for environmental reasons. This book focuses on water drainage design at the building scale, it does not cover the disposal of industrial effluence or municipal transmission, treatment and disposal of wastewater. What is important to note is that there are important design, human/public health and environmental guidelines associated with each stage of the drainage chain. Therefore, different countries have different policies on drainage systems.

For example, in order to safeguard against contamination, infiltration into ground and surface water are not permitted in Austria. Such policies are understandable as there are continuous risks of man-made substances and other pollutants contaminating water sources. The impact of long-term accumulations of these pollutants and contaminants in water systems or the ramifications for the food chain and human health are still not fully understood. However, if most countries banned infiltration, the existing wastewater treatment infrastructure would not be able to cope without considerable investment in new treatment processes and plant solutions.

Other countries have regulations that limit the infiltration and discharge of wastewater into natural systems. Concerning buildings, such building regulations or codes tend to set minimum specifications that will ensure healthy and safe buildings. However, such regulations can be prescriptive and limit innovation. Hence, in a number of countries, the actual regulations can be rather vague, including such words as 'adequate', 'sufficient', 'efficient', 'safe', 'comfortable', 'clean', 'hygienic'; all of which are rather subjective. Such generalized requirements are normally accompanied by examples of how to achieve the actual regulations. These are detailed in guidance documents that can be updated more easily and regularly than statutory legislation. Such documents often 'call-up' national, continental or international standards along with technical documents from industry to provide detailed specifications and rules.

An example is the building regulations of the regions of England, the largest country within the United Kingdom. (The building regulations for Wales and Northern Ireland tend to be similar to those of England, but the regulations in Scotland are based on a different legal system that can be more prescriptive). The Building Regulations (England) Part H: Drainage and Waste Disposal stipulates that an adequate drainage system should be provided to carry foul water from appliances in a building to a public or private sewer, septic tank or cesspool. Any septic tank should be correctly designed, built and located, and all paved areas should be adequately drained to avoid surface water problems. It also makes similar stipulations for the conveyance of rainwater from the roof of a building to a soakaway, an infiltration system, a water course or a sewer, except if the rainwater is being collected for reuse. As much as is feasible, rainwater drainage should be separated from foul water discharge. Only greywater from personal use and clothes washing can be collected for reuse in buildings. All drainage systems should have adequate access for continuation and maintenance, and should be located such that it does not harm people's health.

A number of international standards that emphasize the link between water supply and drainage systems are currently under preparation.

The approach taken by the ISO committee is identified in the extract quoted below from the current ISO 24511:2007 (Activities relating to drinking water and wastewater services — Guidelines for the management of wastewater utilities and for the assessment of wastewater services). This document sets out the role of drain and sewer systems and wastewater treatment and disposal:

*Wastewater systems are built and operated mainly to protect public health and the **environment**. The type of **wastewater system** needs to be chosen and adapted in context with the density of the population, climatic conditions, environmental requirements for treatment and the technical/socio-economical ability of the **responsible body** to implement it, operate it and maintain it. It needs to be cost effective and sustainable, as well as permitting phased development to overcome the financial constraints while not compromising the stated objectives.*

*Operationally, the broad objectives of a utility are to provide **wastewater** collection **services** on a continuous or at least intermittent basis (depending on the service mechanism chosen), meeting the related capacity **requirements**. Methods of **wastewater** treatment and/or disposal need to correspond to the chosen collection system.*

*Appropriately, treated **wastewater** is eventually returned to the **environment** and can have significant impact on both quantity and **quality** of natural water resources.*

*Effective and safe management of **residues** resulting from **wastewater** treatment, including their final disposal or reuse, is becoming increasingly important due to concerns about both environmental protection and resource conservation.*

*Since it often has a lifetime stretching over several human generations, **wastewater infrastructure** needs to demonstrate intergenerational equity. Consequently, a **wastewater utility**, regardless of ownership, is public in nature and will be subject to public scrutiny and policy. Other criteria, such as cost/**affordability** and **service** sustainability, are addressed in appropriate clauses of this International Standard.*

To summarize, wastewater and drainage are not isolated. They are part of a global water cycle that until recently had generally been ignored. Historic drainage system designs tended to focus on managing local wastewater issues and assumed that the vastness of the planet's oceans and soil could cope with whatever was disposed into them. The truth that most resources are finite and that an increasing population uses more resources has led to a re-evaluation of attitudes to waste and our stewardship of the planet. Local regulations are becoming more consistent across the planet and policies and strategies are being put in place internationally. Drainage is no longer simply a disposal route, but a conduit for different resources that can be utilized, reused and sensibly treated before re-entering the global water cycle. Perhaps we should reclassify 'wastewater' as 'used water', which better conveys that there should be no such thing as water that should be wasted.

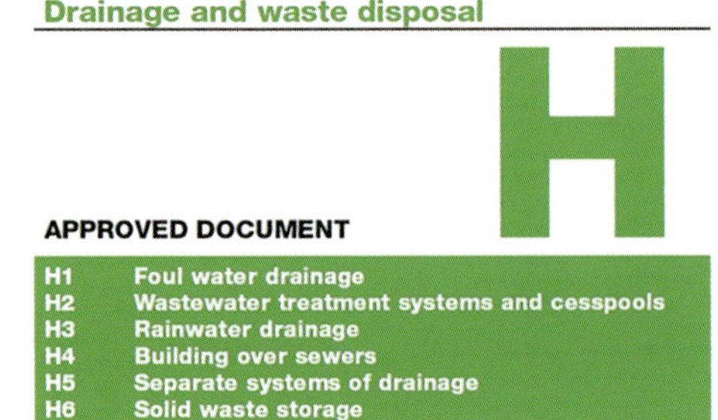

Fig. 0.11 The Building Regulations (England) Part H: Drainage and Waste Disposal.

Chapter 1

WASTEWATER DRAINAGE SYSTEMS

This chapter discusses the evolution of site-based and municipal drainage systems and the importance of good drainage for public health and well-being. It also highlights the importance of good drainage design and the choice of appropriate methods for achieving both water efficiency and resilience. The second part of the chapter will present the main terminology, criteria and factors that inform 'traditional' drainage design in and around buildings and the built environment.

Here, building drainage system types and design considerations are discussed with particular focus on the collection and transmission of rainwater, foul and wastewater from buildings; commonly referred to as sewer systems. In the UK, sewer systems can be private and public, or separate and combined.

Useful standards and guidance documents include:

- BS EN 752 (2017) Drain and sewer systems outside buildings. Sewer system management (formerly seven parts, now only one part)
- BS EN 12056 Part 3 Gravity drainage systems inside buildings – Roof drainage, layout and calculation (although title is for 'inside buildings' the scope includes rainwater drainage attached to a building)
- EN 16933-2 (2018) Drain and sewer systems outside buildings – Design Part 2: Hydraulic design
- BS EN 1610 Construction and testing of drains and sewers

- BS EN 16932-1 (2018) Drain and sewer systems outside buildings. Pumping systems (In 3 parts including Vacuum systems and Positive pressure systems)
- BS EN 295 (2013) Vitrified clay pipe systems for drains and sewers (Currently in seven parts)
- BS 65 (1991) Specification for vitrified clay pipes, fittings and ducts, flexible mechanical joints for use solely with surface water pipes and fittings
- BS 4660 (2000) Thermoplastics ancillary fittings of nominal sizes 110 and 160 for below-ground gravity drainage and sewerage
- CIBSE Guide G (current version): Public Health & Plumbing Engineering.

The EN 752 is a particularly useful reference. It includes a helpful diagram of the various drainage systems that will be found in a typical urban water catchment e.g. river basin. It shows how the different systems are interconnected and can all impact on local water bodies. It differs from previous drain and sewer standards in that environmental issues are foremost and reference is now given to important European directives such as the:

- EU Urban Wastewater Treatment Directive (91/271/EEC)
- EU Bathing Waters Directive (2006/7/EC)
- EU Groundwater Directive (2006/118/EC)
- EU Shellfish Waters Directive (2006/113/EC), and the
- EU Floods Directive (2007/60/EC).

Although the design of a drain or sewer system needs to take into account the anticipated loading, EN 752 also now requires the design to be future-proof by making provisions for:

- increased flows due to new developments;
- reduced flows due to changes in the development and changes in water use;
- increased flows resulting from changes to existing developments;
- reduced hydraulic capacity resulting from increased hydraulic pipeline roughness (e.g. due to deterioration) in the drains or sewers or wear in pumps;
- reduced hydraulic capacity due to build-up of sediments in sewers.

Private and Public Sewer Systems

Generally, when drains and sewers are given ownership (private or public), this means that they are piped conduits. Although the terms public and private drains and sewers are used around the world, they can have very different practical and legal meanings. Drains and sewers may be installed in the wealthier parts of the towns if treatment facilities have been developed and there are continuous water supplies, but installations in poorer areas may be limited by water supplies, topography and the high densities of population. In areas where pit latrines predominate, sewerage systems are less common. In such situations, a drain may simply be a ditch and to a nearby water body, e.g. a river.

A private sewer system consists of drainage situated within the boundary of private land or property. This means that the land or property owner is responsible for repair and maintenance. A private sewer can serve multiple housing units, e.g. apartments or buildings within the private boundary whose system of drains then combine to single outlet to a public sewer system.

Public or adopted sewers refers to drains and pipelines outside the property boundary. These are maintained by the local authority or municipality. In England and Wales, The Water Industry (Schemes for Adoption of Private Sewers) Regulations 2011 states that with some exceptions, all privately owned sewers and lateral drains (see definitions) that communicate with or drains to an existing public sewer as at 1 July 2011 will become the responsibility of the sewerage undertaker – normally the water and sewerage company for the area. However, property owners will still be responsible for the sections of pipe between their property/building and the transferred private sewer or lateral drain. This rule also applies to surface water sewers except those that drain to a water body or outlet other than the public sewer (see Fig. 1.1).

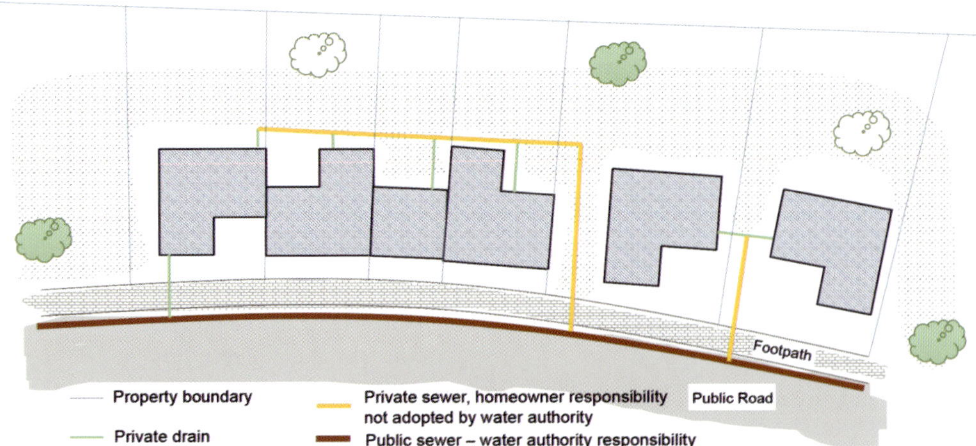

Fig. 1.1 Private Sewer Adoption since 2011. (Adapted from: Defra (2011) The Private Sewer Transfer Regulations: Provisional non-statutory guidance on private sewers transfer regulations. Crown Copyright. June 2011)

Separate and Combined Sewer Systems

Drainage systems consist of above and below-ground networks of pipes designed to convey foul, surface and rainwater to the sewers. These are typically situated to run alongside or under road networks. The below-ground drainage systems can be separate or combined.

A separate drainage system involves using one sewer system to remove rainwater collected from the roof and other surfaces to a public sewer network, and another to remove the foul and wastewater to a different public sewer network. BS EN 1085:2007 Wastewater treatment. Vocabulary [definition 2120] defines it as two pipelines, one carrying foul wastewater and the other surface water.

A combined system is where both the rain and foul water are removed via the same public sewer network. Historically and pre-1920s in the UK, a combined sewer was more common. However, there are advantages to a separate sewer system that includes: managing rainwater peak flow – reducing the risks of overflows and backflows of the system; reducing treatment loads since surface water is less contaminated than foul water; and reducing the risk of sewage contamination of water bodies. The situation in most other industrialized countries around the world is similar. Some places, such as Washington in the USA, have tried to separate existing combined sewers, but the cost and disruption is very high.

The trend towards separated systems with sewer systems has increased over the last few decades, as there are more concerns over discharges from combined sewer overflows (CSOs) and a general desire to treat foul water

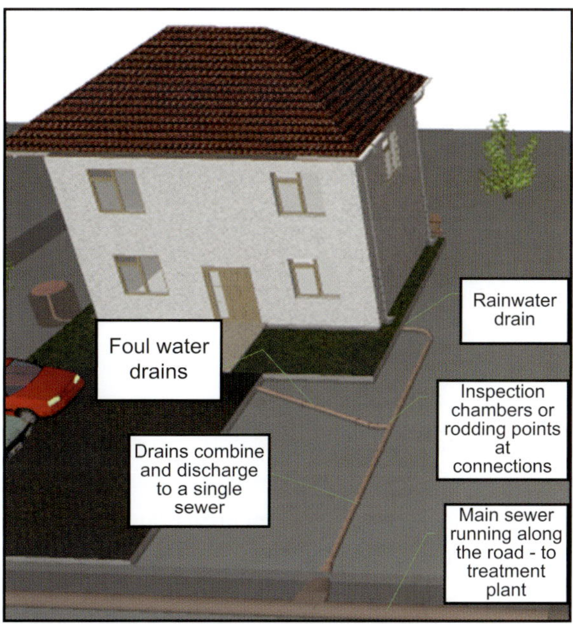

Fig. 1.2a Combined drainage systems.

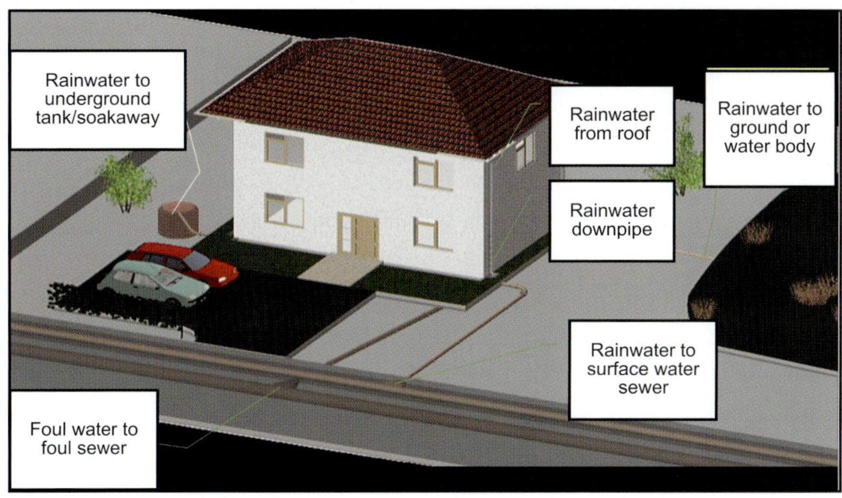

Fig. 1.2b Different options for separate drainage systems.

separately to stormwater. However, there is less prevalence of separate drainage systems within buildings, where greywater and blackwater could be separated, as the cleansing effect of flushes from WCs (blackwater) helps to minimize the deposits from surfactant using appliances such as sinks, baths and showers.

However, although combined systems can be connected together below ground, many systems exist where rainwater and foul water are combined above ground; for instance using hoppers or branch connections.

Other Drainage Systems

The current hierarchy for disposal of foul water from a building, in England, is:
a) A public sewer
b) A private sewer
c) A septic tank or other wastewater treatment system
d) A cesspool
The criterion for selecting the option is practicability. Therefore, an on-site treatment system is currently prohibited in the UK unless it is impracticable to connect to a sewer. In the future, and in other countries, this hierarchy may change as the water discharged from a building becomes seen as a resource more than waste.

Vacuum or Pumped Systems

In areas where the ground is flat, drains may start shallow but can become very deep if there is a considerable length of run. In such situations, alternatives to gravity drainage may become necessary. The main options are pumped systems; pressure or vacuum.

With vacuum systems, the wastewater within a building may be collected by vacuum or by gravity, or a combination. The collected wastewater is directed to an interface valve chamber, where the water accumulates. When sufficient water has been collected, detected by level sensors, the interface valve will open and discharge the water into the vacuum sewer pipes. The vacuum

pipelines run in a shallow sawtooth profile and wastewater is transported in pockets that form at the low points of the pipes.

Pressure systems use full pipes that can be run at any length or height that are within the capability of the pump used. In some complex drainage systems, parts may be vacuum- or pressure-driven. The use of any pumped system requires a power supply, so it is also normal to provide back-up power and standby pumps to improve the reliability of the system and prevent system failure.

On-site Solutions

In situations where the building is remote from any established drain or sewer system, an on-site collection or treatment system may be used. Collection systems rely on tankers regularly emptying receiving containers for the effluent. A treatment system can reduce the level of pollutants in wastewater to a level that is safe to discharge, reuse or even drink; depending upon the system used and the local regulations and practices. The level of treatment that is required impacts the: cost, footprint, energy consumption and maintenance of any system.

Cesspool

The most common type of receiver is a cesspool. These are sealed tanks, usually installed below ground, that prevent wastewater leaking into the environment. Access to the cesspool needs to be created and maintained so that routine emptying can take place. Hence the cesspool needs to be sited no more than 30m from vehicle access, unless longer emptying pipes are available from local emptying companies. The location also should be at least 7m from a habitable building and, if possible, downslope from the building. As cesspools do not perform any treatment, they should only be used when no other option is possible.

Septic Tank

This is the most common on-site system used for build-

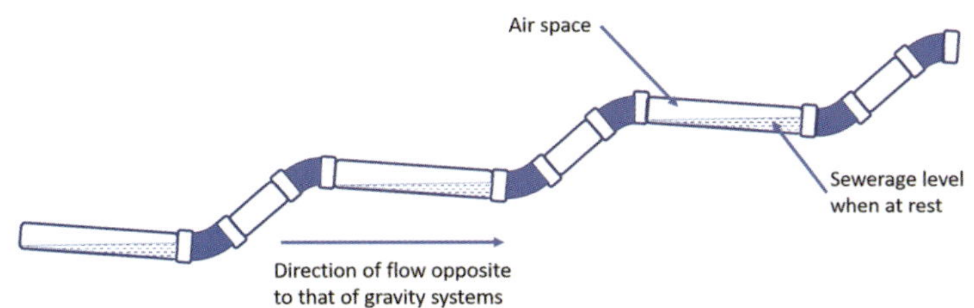

Fig. 1.3a Vacuum sewer 'sawtooth' profile; b. a typical collection unit. (Source: Adapted from Naret, J. (2007) Fig. 6 in: Mohr, M., Iden, J. and Beckett, M. (December 2016) Guideline: Vacuum sewer systems, Fraunhofer-Institut für Grenzflächenund Bioverfahrenstechnik IGB, Stuttgart)

ings. This receiver is a flow-through device that provides primary and some secondary levels of treatment. The shape of the container encourages settlement of solids and flotation of low-density material. The settled liquid then passes into subsequent chamber or chambers for further treatment before discharge. The discharge from a septic tank (treated effluent) is normally directed to a drainage field or other infiltration system where additional treatment is provided as the water is cleaned by the action of the soil.

In the UK, any septic tank or packaged treatment plant needs to meet the requirements of EN 12566 and the drainage field should comply with BS 6297:2007. All septic tanks and small wastewater treatment plants (packaged plants) need to be de-sludged annually so that their performance can be maintained.

While septic tanks are simple passive containers, most alternatives use electrical power in various ways to introduce more oxygen into the wastewater, so they are not immune to power cuts. Such systems are sometimes called packaged treatment plants.

Bio-discs: These devices are normally partially buried and contain the wastewater in an open-topped trough into which a number of discs have around a third of their areas immersed in the slowly flowing wastewater. The discs are attached to a central axis that slowly rotates the discs. The discs contain corrugations that increase the surface area for the wastewater to produce a film of bacteria that can react with the air. The oxidation reactions help break down the wastewater.

Aeration systems: Submerged outlets introduce air that is blown to the outlets using electric fans and pumps. The

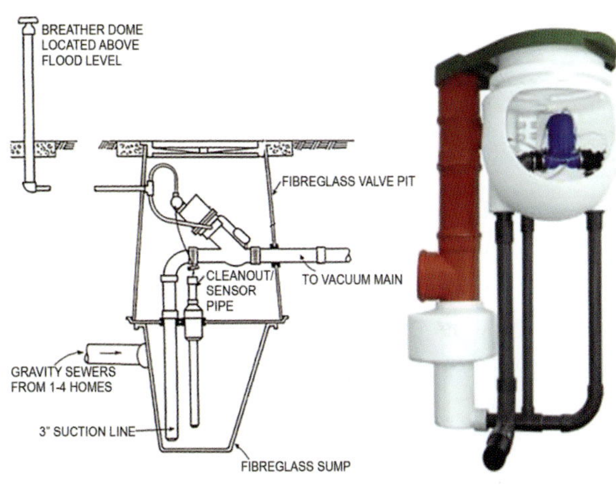

Fig. 1.3b
L: Schematic layout of a collection pit with vacuum valve: R: Prefabricated collection pit.

air introduces more oxygen to the wastewater, enabling the breakdown of the wastewater. Aeration systems can be relatively noisy.

Pumped systems: Advance systems may use membranes in various locations to produce higher-quality final effluent. Pumps are needed to force the wastewater through the membrane. A control system will also be required for periodic backwashing to clean the membranes. Although such systems are expensive and need power to run, they can be very compact.

Condensate systems: Small soakaways can also be used for condensate drainage. In this case, the small dispersion unit is filled with calcium carbonate to help neutralize the acidity of the condensate before infiltration. Due to the small volumes of water that are discharged at any time, the soil type does not significantly influence the

Fig. 1.4 Condensate drain from a boiler, white plastic, discharging through a valve to the wastewater branch.

Fig. 1.5a Example of boiler condensate drainage to a WC branch within an inspection chamber with cover open.

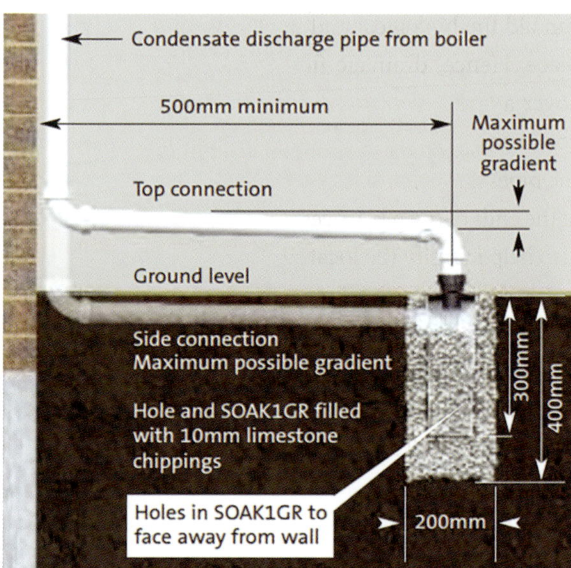

Fig. 1.5b. Condensate drain soakaway.
(Source: McAlpine https://mcalpineplumbing.com/sites/
default/files/uploads/new_condensate_brochure.pdf)

infiltration rate.

Condensate drainage may also be into a foul water drain. If this is to be done, the connection needs to be near an outlet that will provide frequent dilution, such as from a WC branch. The condensate drain will need to be trapped, but this may be achieved within the boiler, so the manufacturer's instructions will need to be read.

Reed Beds and Constructed Wetlands

These systems can be used for wastewater as well as stormwater. The difference is mainly apparent from the size. Wastewater constructed wetland systems tend to be larger with a much greater land-take than stormwater ones, where more parts will provide infiltration to ground.

For wastewater treatment, the constructed wetland will normally follow a primary treatment stage in which heavy solids settle and light materials float. Although reed beds and constructed wetlands can provide some biological treatment, the main mechanism is physical due to the filtering aspects of the sand layers within the structures. The main advantage of such systems is that they require no power and can enhance environments. However, the land area required for such systems can be prohibitive and the excavation and construction can be very expensive.

The final stage of a constructed wetland may be a discharge to a watercourse or water body, if one is nearby, or to the soil via a conventional drainage field. The soil provides an additional level of treatment that further polishes the treated effluent. Most treatment takes place in the top few metres of the soil, where air is present and can aid the biological and chemical reactions that take place. Hence, drainage fields tend to be shallow and cover a large area of ground. With any drainage field construction, it is important to remember that the air in the pipes needs to be displaced by any water; ventilation at the ends of the pipes needs to be provided. These vents also help identify the location of the field to subsequent users of the site and can provide access points for maintenance. Although drainage fields allow the treated wastewater to soak into the ground, they are not soakaways in the normal definition as they provide long-term treatment as well as infiltration.

Soakaway

These are chambers, below ground, that are designed to leak water; they are used to disperse surface and stormwater where their discharge to the sewer is unwanted or not possible. As infiltration is an important component of Sustainable Drainage Systems (SuDS), soakaways are an important element in this method (see SuDS section later in book). Soakaways can deal with stormwater at source, but they are generally not suitable for clay soils. Although new soakaways are normally specified to be at least 5m from a building, many existing soakaways can be found far closer, especially in small plots where to locate a soakaway beyond 5m would place it in another property or highway.

The hierarchy of disposal of surface water, in England, is inverse to the foul water discharge and is currently:

a) To an adequate soakaway or other adequate infiltration system, or

b) A watercourse, or

c) A sewer,

The criterion for not using any of the suggestions is practicability, so a sewer is only to be used as a last option. The term 'adequate' is popular in regulations, but in this instance, it implies that the infiltration system will cope with the rainfall from a storm with a return period of ten years.

Fig. 1.6 A reed bed wastewater treatment system under construction. Note the bark rings to the right that provide the initial filtering, followed down the slope by two vertical reed beds and finally, to the left, a horizontal reed bed. This is to serve a large farm house but has occupied a large area of land.

Fig. 1.7 A well-maintained and constructed vertical reed bed serving a visitor centre, showing the ventilation and access pipes.

Site Considerations

When it is necessary to discharge treated wastewater to ground the local ground conditions will need to be assessed. BS 6297 includes a series of tables that will help identify a good or bad site from simple visual indications, such as the types of plants thriving in the area. In addition, the maximum ground water table level will need to be determined and this should be at least 1m below the lowest level of the proposed drainage field.

This is often determined by digging a number of test holes on the site; any that immediately fill with water is an indication of an unsuitable location.

Test holes are small holes that are used to determine the infiltration rate for the ground. They can be affected by various soil features such as stones, rocks, clays and sands; hence infiltration tests can give a wide range of results. To try to improve the accuracy of infiltration tests many have been devised, but they all have limitations. Most tests will determine the vertical percolation rate (V_p number). However, most infiltration systems will also have a horizontal component of percolation, H_p, which is rarely measured with simple tests. So, the results of an infiltration test will indicate the performance of the soil at that location, but the infiltration can vary by weather, season and any preceding discharges.

A percolation test needs to be carried out on a representative section of the receiving soil. If the soakaway is serving an area of less than 100m^2, a simple test on a small hole will suffice. However, if the area is much greater much more complicated and expensive tests may be needed to determine the infiltration rate for the whole infiltration site. Once infiltration rates are determined a calculation can be made to work out the outflow rate and compared with the specified inflow rate to determine the storage volume that will be required.

A typical infiltration test would be similar to the BS 6297-based test outlined here:

I.   A square hole about 300mm deep below the lowest level of the soakaway or infiltration device.

II. Fill the hole with water and let it all soak away. If it drains away immediately, the soil will be good for infiltration, but not offer much treatment of any contaminants. If the water does not drain away within one day the site will be unsuitable.

III. After the first filling has drained, refill the hole and commence timing.

IV. Once the level has dropped to about 75 per cent and again at 25 per cent, record the times.

V.   The time between 75 per cent and 25 per cent should be divided by the vertical distance between these two locations. This will produce the vertical percolation rate V_p in s/mm.

VI. The whole test should be repeated a number of times in various locations within the proposed site for the infiltration system and the results averaged.

Ideally, the average vertical percolation rate should be between 12 and 100. However, for stormwater a V_p of less than 12 would be acceptable. If it is greater than 100, the required storage may be problematical.

For soakaway design, the vertical percolation rate is a key parameter, but in reality, it is very difficult to determine with any degree of accuracy. A bulk percolation rate determined in situ with a realistic quantity of water is the only real way to determine the percolation rate, but this is often impracticable. Hence, various percolation test methods can be used.

The tube sample method gives a very accurate vertical percolation rate, but it is questionable as to whether or not it is meaningful in all situations. The test pit method has a number of variations, but is essentially a small hole in the ground into which water is poured and the time taken to drain is measured. Depending on the soil type, various errors can be introduced by smearing of the soil, small rocks and stones in the surfaces, etc. Hence, some methods require the insertion of a sleeve to stop side infiltration, others require the sides to be roughened to increase the side infiltration. Some use falling heads, others use reservoirs to produce constant head tests. Different techniques will give different results, so it is important to be consistent with which tests are used and how.

Here are some alterative infiltration methods:

ASTM D3385 – 18 'Standard Test Method for Infiltration Rate of Soils in Field Using Double-Ring Infiltrometer'

ASTM D5093 – 15e1 'Standard Test Method for Field Measurement of Infiltration Rate Using Double-Ring Infiltrometer with Sealed-Inner Ring'

ASTM D6391 – 11 'Standard Test Method for Field Measurement of Hydraulic Conductivity Using Borehole Infiltration'

EN ISO 22282-2:2012. Geotechnical investigation and testing. Geo-hydraulic testing. Water permeability tests in a borehole using open systems

Such tests may also be applied to surfaces that are intended to infiltrate water, such as porous pavements and block paving. Inundation tests can also be used if there is sufficient water available. To determine the volume of water that would flow into the soakaway, the following calculation (from English Building Regulations Approved Document H Paragraphs 3.26-3.29) can be used.

From the vertical infiltration rate, V_p, in seconds/mm, the soil infiltration rate, f, in m/second, can be obtained from:

$$f = \frac{10^{-3}}{3V_p}$$

This is used to calculate the outflow volume, O:

$$O = a_{S50} \times f \times D$$

Where:

a_{S50} = total area of the sides of the storage volume, when filled to 50 per cent of its effective depth (m²)

f = Soil infiltration rate (m/s)

D = Storm duration, in minutes

Alternative infiltration calculation methods are available and the results are often quite different and result in different designs of infiltration systems. However, as infiltration tests are wildly variable in their results, this is not surprising. Results for different tests on the same site can vary by factors of several hundreds and will also vary on a daily basis and much more on a seasonal basis. Infiltration testing does not produce accurate values; it does produce indicative values that are helpful in designing an infiltration system.

Discharges to watercourses and sewers do not require soil investigations or calculations.

Various innovations have advanced soakaway construction over the past twenty to thirty years. The main elements are geotextiles and plastic crates.

Geotextiles are now available in a vast range of specifications and performance. This means that they can be used to: filter fines i.e. small particles, protect structures from point loads, provide a filter medium, provide a waterproof covering, provide a breathable covering that will enhance oxygen transmission, hold structures in place, prevent root or plant ingress, allow root penetration, prevent gas transmission, and various combinations of effects. Hence, when specifying a geotextile it is vital that the full specification and purpose of the material is given.

Geotextiles are now used to help form sand and gravel filters that can provide various types of wastewater treatment, as well as simple infiltration. They are also used to wrap crates that can be used for stormwater attenuation and infiltration. The emphasis on stormwater attenuation is the result of recent flooding events, in various parts of the world, highlighting that traditional rainwater systems cannot cope with high levels of rainfall.

Hence, soakaways are now no longer stand-alone features, but are frequently combined with attenuation devices, rainwater reuse systems and meteorological data to produce active systems that will infiltrate when required, but can also store water when heavy rainfall is not expected. In practice, this means that the soakaways may be smaller than traditional designs as extra storage and flow-limiting devices can reduce the discharge flow rates. Partial infiltration can also be built into some of the storage devices, increasing the options for infiltration routes.

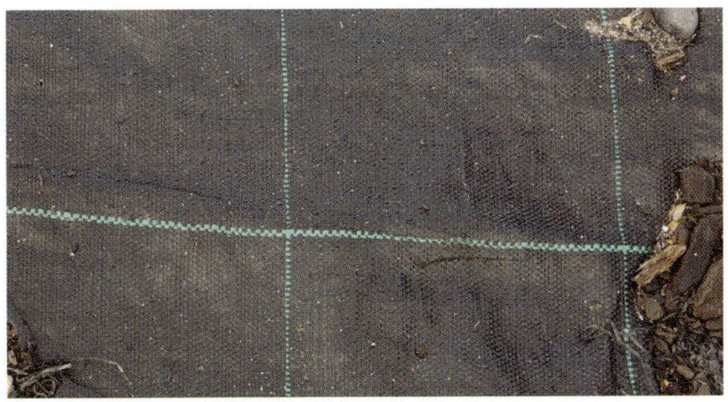

Fig. 1.8 Example of one type of geotextile used for preventing plant or root penetration.

Consent

Irrespective of the on-site solution, there will be a specified quality of water that may be discharged either to a water body or to the soil (especially to avoid contamination of underground aquifers). In the UK, this is specified in the 'discharge consent' or Permit issued by the local environmental agency or authority. A discharge consent may not be required if the volume of wastewater is low. In the UK, the discharge consent is also covered by the rules that sewage must:

1. be domestic in nature, for example from a toilet, bathroom, shower or kitchen of a house, flat or business (such as a pub, hotel or office)
2. not cause pollution.

A permit is required if the discharge is to be more than 5,000l (5m³) a day to a water body.

A permit must be obtained if the discharge is to the soil, if the volume is greater than 2,000l (2m³), or the discharge is to a well, borehole or other deep structure, or it is within a groundwater source protection zone.

A groundwater source protection zone in England is defined as:

- the area around a commercial water supply (used for drinking water or food production) shown on the map of protected zones – check if your discharge is in the inner zone (zone 1) or ask the Environment Agency
- any area within 50m of a private water supply for human consumption – ask your neighbours if they have one and if so how far their spring, well or borehole is from your drainage field

There are other areas where permits may be required, such as ancient woodlands and conservation areas. If in doubt, seek confirmation from the relevant authority before constructing any on-site wastewater discharge facility.

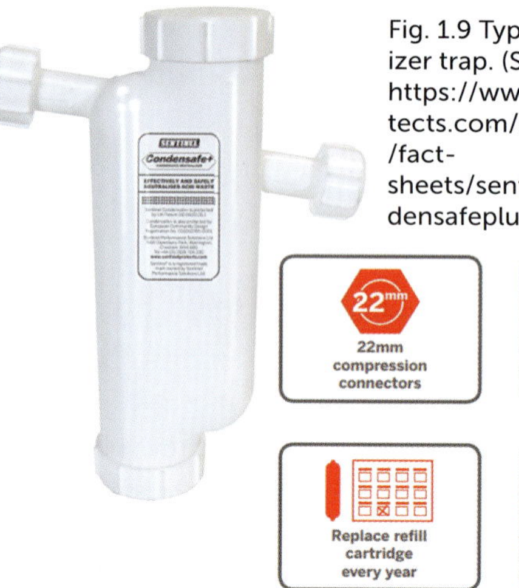

Fig. 1.9 Typical pH neutralizer trap. (Source: https://www.sentinelprotects.com/sites/default/files/fact-sheets/sentinel_pfs_condensafeplus_uk_v2_june_2

Chapter Summary

There is no such thing as a typical drainage system as they come in a wide range of configurations. Some discharge to a sewer, others to a tank or an on-site treatment plant. At building level, rainwater and foul water will normally be collected separately, but outside the building the two systems may be combined. A combined system only uses one pipe, so it is simpler to install, but a separate system that uses two pipes can provide more opportunities for reuse and reduce the number of incidents of foul water being discharged to water bodies, such as rivers.

Although gravity is the most common motive force used to move wastewater, various pumped systems are available. The legislative preferences for the disposal of surface water and foul water are opposites. The preferred route for wastewater is to a sewer, but for surface water it is infiltration. Infiltration to soil is a widely accepted route, but the sites for infiltration need to be carefully assessed and designed.

On-site treatment systems can range from just a collection tank, confusingly known as a cesspool, to through-flow or batch treatment systems that will produce an effluent of a suitable quality to be discharged to the environment.

Chapter 2

FOUL WATER DRAINAGE DESIGN

This section addresses the removal of wastewater and associated solids from the building. It will again cover the standards associated with water drainage design, introduce key design principles and present examples.

Waste or foul water from buildings contain liquid and solid components and foul drainage systems needs to be designed properly to cope with the volume, type and frequency of the wastewater generated.

According to the approved documents to the English Building Regulations Part H1, the performance requirements for foul drainage would be considered met if the following criteria are met:

- Conveys the flow of foul water to a foul water outfall (a foul or combined sewer, a cesspool, septic tank or holding tank)
- Minimizes the risk of blockage or leakage
- Prevents foul air from the drainage system from entering the building under working conditions
- Is ventilated
- Is accessible for clearing blockages; and
- Does not increase the vulnerability of the building to flooding.

Drainage Location

The choice of location of drainage stacks is dictated by the location of appliances and or the existing drain runs. Although it is not always possible, it is not good practice to run below-ground drains beneath buildings (due to structural problems and the difficulty of access if there is a problem). Drains are better if run around the outside edge of a building. This makes external above-ground drainage an easy and simpler option. In France, pipework is often external to the building, but not obtrusive. The merits of different locations for above-ground drainage are summarized in Table 2.1.

Drainage Principles

A well-designed and installed drainage system helps to avoid problems such as sewer backflow and flooding, especially during storms. It also prevents standing foul water, foul odour, insects and rodent infestations, contaminated land and soil, among other issues. So, this section covers some important principles and components of good drainage design.

There are two common types of drainage systems; pumped (positive pressure or vacuum) or gravity. In the UK, most sewers still run gravity systems whilst some cities have pumped, or combined pumped and gravity, systems. The trend towards pumped systems at the city or municipal level is typically to accommodate difficult terrain or sometimes to facilitate heat re-

Main location	Outside of building	Inside buildings	Within accessible service ducts
Pros	Excellent accessibility for maintenance and modifications. Clearly show the quality of the install-ation Can be made into interesting visual architectural features.	Protection from frost and freezing. Much reduced working at height needs for installation and maintenance.	Protection from frost and freezing. Reasonable accessibility for maintenance and modifications. Minimal working at height needs for installation and maintenance. Minimal disruption to inhabitants during maintenance.
Cons	Can be ugly and unsightly. Risk of wastewater, including condensate water, freezing in cold conditions. May require scaffolding and working at height to access parts of the system. Can be affected by plants such as ivy that could weaken fixings and damage joints.	Space needed within the living space and can create obstructions within rooms. Accessibility for maint-enance and modific-ation can be very restricted and difficult. Sound transmission may be a nuisance. Special sound absorb-ing pipework may be specified. Fire stopping needed between floors. Maintenance can be disruptive for inhabitants.	Space requirement for an accessible duct can be prohibitive. Sound transmission may be a nuisance. Additional fire stopping, or other methods, needed to prevent spread of fire.

Table 2.1 Above-ground drainage location summary.

covery and energy generation from sewer waste. The sewers normally lead to municipal wastewater treatment plants where the water can be cleaned before discharge and any solid organic matter extracted as sludge. Sludge and related products can then be reused by agriculture and other sectors, e.g. as fertilizers or for fuel.

Foul drainage systems outside the building are typically conveyed via gravity but pumped systems may also be used; this book focuses on gravity systems only. The main standard for the design of such system is the BS EN 752:2017 Drain and sewer systems outside buildings – sewer system management. In other European countries, the EN is available in German or French translations. The Building Regulations (England and Wales) also cite the older, multi-part, EN 752: 2008 version. This is a common issue in most countries where legislation tends to only specify dated publications. Hence, an updated standard may not be called up in legislation or its guidance for many years, until the base legislation is also revised or amended. The European Standard EN 752 originally provided the framework for the design, construction, operation and maintenance of sewer systems outside buildings. However, the current version now only covers sewer system management.

The following design considerations are important when designing the external drainage system for a building:

- The **capacity** must be large enough to carry the expected flow at any point. Capacity is determined by the expected volume of discharge, the flow rate, gradient and layout of the system.
- Drains should be **positioned** in a simple, straight manner. Changes in direction should be kept to a minimum.
- Gentle bends should only be introduced if it is unavoidable. If used:
  - They should be as close as possible to, or inside, an inspection chamber
  - They should also be close to the foot of discharge pipes
  - The pipe diameter should have as large a radius as possible
- Drains located close to a building or foundations should be **protected** against settlement. **Beddings**, e.g. sand and/or concrete, should be provided too if they are situated close to buildings, or run though or under walls, foundations and other structural elements. Where the pipe runs through walls, at least 50mm of clear vermin-proof space should be provided around it. The run of the pipe should be a maximum 600mm in length, and have joints within 150mm of the wall.
- **Access** points should be sufficient and suitably located to enable maintenance and remove any blockage that may occur. Access should be

Fig. 2.1 Excavation and installation of drains.

provided at:
- the head of each drain run
- at a bend or change in gradient
change in pipe size, and
- at a junction
- Also, structural damage and problems with access and continuation should be avoided by locating sewers away from sites for future development.

Groundworks

Once the wastewater leaves the drainage stacks and enters the near-horizontal drains, on the way to the sewers or local treatment plant, there are more choices. The type of buried drainage system that is used will depend on a number of factors: topography, soil type, location of nearby buildings or trees, soil contamination levels, stability of the ground, nearby waterways and water supplies (e.g. wells), location of any soakaways, risk of flooding, opportunities for reuse, cost and availability.

In hilly areas or where the sewers run deep, excavation can be very difficult and costly. Building deep inspection chambers or access chambers can also be difficult and increases the risk and complexity of maintaining the system due to working at height and confined space regulations. Some alternative systems use high-level access chambers that are located just below finished ground level and have 45 degree outlet pipes that drop down to the drain or sewer.

In low-lying areas, small pumping stations may be required. These will need good access for maintenance and may need hoists or cranes to enable pump and valve servicing and inspection. They may also need to be acoustically and visually screened. As with most wastewater systems, such pumps should have an equivalent of at least twenty-four hours of storage capacity, so that in the event of failure the drainage system can still be used for a limited time.

In addition to low-lying areas, large drainage basins may also need pumped systems. For example, in Lon-

Fig. 2.3 Sewage pumping station at London's Crossness was used from 1864 to 1956. (Source: http://hotten.net/open/pages_large/families/knight/lrk_history/crossness/01.htm)

don the sewers tend to run eastwards towards the North Sea. As the core area of London is about 40 miles wide, a single fall sewer would end up being impossibly deep at its outflow. Hence, a sawtooth layout of sewers is used with pumping stations every few miles to enable reasonable falls to be produced without excessive tunnelling through the clay, sand and chalk substrates of the Thames basin. The pumping stations were often elaborate and inspiring, as shown in Fig. 2.3.

Where soil is not stable and liable to some movement, flexible joints or pipes should be specified. However, if flexible pipes are used the gradients should be steep enough to prevent the formation of back flows (negative gradients). So, casing in concrete may be considered. Rocker pipes (short lengths of rigid pipes with flexible couplings at each end) are often used where drainage pipes penetrate foundations and walls. If these are used with pipes that are laid in gravel backfill, boards that are attached to the pipes will be needed on either side of the wall to prevent loss of backfill and to inhibit any rodent movement through the void around the pipe.

Where the soil is very rocky or comprises stone,

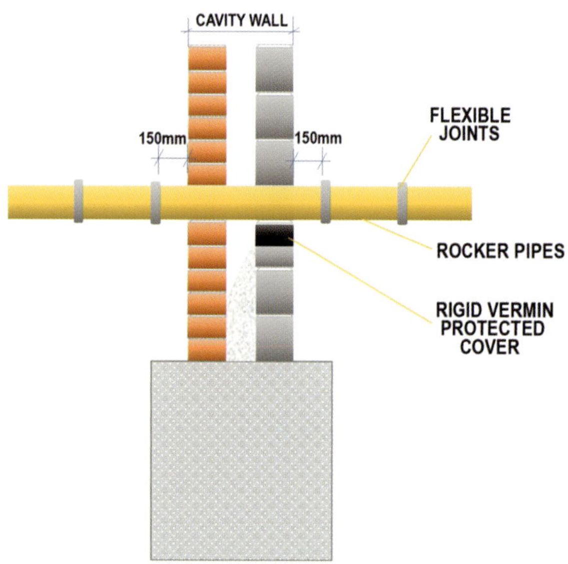

Fig. 2.2. Interface between drainage and building structure – using rocker pipes.

deep drains will be very difficult and expensive to install. In such circumstances alternatives that do not require falls may be considered. A pressure drainage system, in which wastewater is pumped along a pipe, is a simple solution but needs to be sized so that the wastewater is not left in the pipe long enough for it to start breaking down and releasing gases such as hydrogen sulphide.

Trees in themselves are not a problem for foul drainage systems, but penetrating tree roots tend to 'seek' moisture to feed the tree. They can easily multiply once they enter a drain, through a crack or faulty joint or a gap in an inspection chamber, due to the regular supply of water. Very quickly a drain can become

Fig. 2.4 Visible impact of trees on road gutter and pavement.

blocked by tree roots. Tree root cutting tools are available, but doing this can be expensive and disruptive. Tree preservation orders (TPO) in the UK also means that removing the tree during construction is not always an option. Therefore, the design solution is to route or reroute the drain to avoid this problem. Also, drains should be designed with minimal joints and trenches. They should be protected with growth-protective membranes and materials such as concrete where beneficial. The pipe could also be sleeved using an epoxy or plastic liner where there are no nearby connections, but if there are many connections the risk of penetration increases due to the difficulty of making lateral joints in sleeves.

Normally, the main concern with wastewater pipes is that they are suitable for conveying the type of anticipated effluent. However, if the pipes are to be run through brown-field sites, contaminated areas or near areas that experience substantial vibrations and land movement, e.g. near railway lines, then further protection of the pipe and its content needs to be considered. Groundworks for below-ground drainage systems consist of trenches laid in the correct positions, using appropriate gradients and beddings as well as protecting such trenches and embedded pipes from damage. The key design points are summarized as follows:

- Underground drains are positioned in trenches
- This serves to protect them from movement, damage and intrusion
- The depth of the trench is determined by:
  - The point at which the drain enters the public sewer
  - The layout, positioning of the drain
  - Site, soil conditions and slope
  - Required depth of pipe cover to meet protection requirements, e.g. from traffic, foundation, etc.

The digging of trenches and laying of backfill is generally done by hand or machines. The resulting trench will only be accurate in fall within the granularity of the material that is being excavated. Laser alignment tools are available for laying pipes, but usually a simple level and ruler arrangement will produce accurate enough gradients. Precise gradients are impossible to achieve and should not be specified. Once the drain or sewer is laid it will be subject to settlement, so small backfalls are to be expected. As long as the backfalls are small in length and percentage of the drain run, there should not be a problem in normal use. However, to avoid significant backfalls or displacements of pipe joints, mechanical plant should not be parked or traversed over freshly laid pipework.

Drainage Components

Drainage systems comprise a number of components, such as drainage trenches, pipes, connections, gullies and access points (Fig. 2.5).

Drainpipes

Drains are typically pipes laid directly into dug or pre-fabricated trenches, which can be up to about 10m deep. The two main materials used for drainage pipes are clay and plastic. Other materials that have been used in sewer networks internationally include concrete, asbestos cement and iron. Modern sewers now mostly use plastic pipes. These pipes can be rigid or flexible, with the latter mostly used at joints or connections. Table 2.2 summarizes the applicable UK standards for each material type with particular emphasis on clay and plastic.

Rigid and flexible pipes can also be used in drainage systems. Rigid pipes rely on their shape, thickness and joints to withstand anticipated dead and live loads. However, the quality of bedding and backfill (see next session) becomes important to maintain pipe structural integrity and avoid damage.

On the other hand, flexible pipes have less material stiffness and deform to cope with, and transfer, imposed loads to the surrounding bedding material without cracking. So, the embedding material is more important, and well graded, granular material is typically used. Any cost saving by specifying flexible pipework can be negated when the additional cost of backfilling and any concrete bedding or covering is taken into account. Flexible pipes are preferred in unstable ground conditions that experience significant movement and deformation.

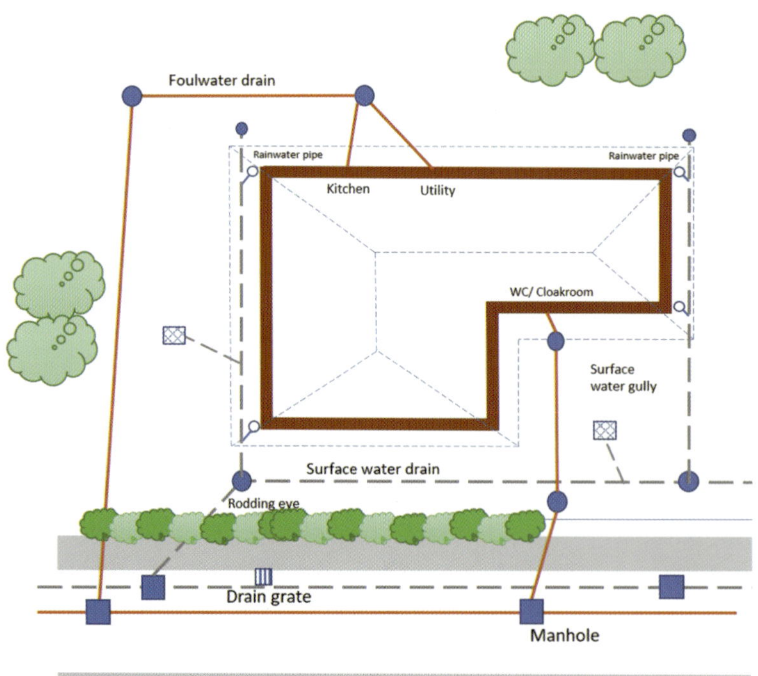

Fig. 2.5a Site drainage plan showing key drainage components.

Fig. 2.5b–e Flexible plastic pipes installed in trenches using pea shingle bedding, prior to backfilling in soil. The installation shown in Fig. 2.5e may need to be bedded in concrete as it is close to the building's foundations.

Material	British Standard	Advantages	Disadvantage
Rigid Pipes			
Vitrified Clay	BS 65 specification for vitrified clay pipes, fittings and ducts, also flexible mechanical joints for use solely with surface water pipes and fittings. BS EN 295 vitrified clay pipe systems for drains and sewers (various parts).	Vitrified clay can be stronger than plastic and less prone to deformation once buried. This places fewer requirements on the bedding material. Depending on how it is sourced, some clay ware has lower embodied energy than plastic. It is resistant to chemical attack, attack by vermin and the use of high pressure drain cleaning.	Can fracture if there is excessive ground movement or imposed loads.
Concrete	BS 5911 concrete pipes and ancillary concrete products (Various parts). BS EN 1917 Concrete inspection chambers and inspection chambers, unreinforced, steel fibre and reinforced.	Mainly used in civil engineer projects and pipejacking, but common for precast inspection chambers and chambers as well as soakaways. The inherent strength and stability requires minimal backfilling.	Heavy, so need mechanical help in positioning the pipes, etc.
Grey Iron	BS 437 specification for cast iron drainpipes, fittings and their joints for socketed and socket-less systems.	This standard has been in use, in various forms since the 1930s, so it is well established and the product well known.	Robust, but heavy fittings require mechanical joints that often involve spanners and bolts. Previously, they were calked with lead and cement to join so joints were slow to make.
Ductile Iron	BS EN 598 ductile iron pipes, fittings, accessories and their joints for sewerage applications. A new version of the standard has been under preparation since 2017 with a different title 'Coated and lined ductile iron pipes, fittings and their joints for sewerage and drainage applications.'	Lighter in weight to conventional cast iron, but similarly robust.	

Table 2.2 Materials for below-ground drainage pipes and connections in dwellings.

Material	British Standard	Advantages	Disadvantage
Flexible Pipes			
UPVC	BS EN 1401-1 plastic piping systems for non-pressure underground drainage and sewerage. Unplasticized poly vinyl chloride (PVC-U).	Plastic (e.g. uPVC) pipes are lightweight and easier to handle on site. Other advantages include flexibility of shape, size and form that plastic affords for the design and installation of drain systems. Different plastics have different chemical resistances to both internal and external substances.	Disadvantage: they are susceptible to deformation when buried. Therefore, plastic pipes must be bedded in pea shingle or selected small gravel to protect the pipe. Different plastics have different chemical resistances to both internal and external substances.
PP	BS EN 1852-1 plastics piping systems for non-pressure underground drainage and sewerage. Polypropylene (PP).	Structured wall pipes are even lighter than solid wall plastics pipes, they are available with either smooth surfaces or corrugated surfaces on either the outside or inside as used in different applications of perforated pipes such as irrigation or infiltration.	There are concerns that structured wall pipes are more prone to damage by rough backfilling or rodding.
Structure walled plastic pipes	BS EN 13476-1 plastics piping systems for non-pressure underground drainage and sewerage. Structured-wall piping systems of plasticized poly (vinyl chloride) (PVC-U), polypropylene (PP) and polyethylene (PE).		

(Adapted from Building Regulations (2010) Approved Document H with added details for commonly used materials.)

Backfill and Bedding

Backfill and bedding are essential to provide protection for pipes. The choice of bedding depends on:
- The type, size and strength of pipe
- Depth of the trench and cover to the pipe.

See the AD-H (English version) for limits of cover for different types of pipes and the minimum bedding requirements (excerpts in Fig. 2.7).

If a drain is located close to a foundation:
- Within 1m: The trench should be filled with concrete up to the lowest level of the foundation
- More than 1m: The trench should be filled with concrete to a level below the lowest level of the building, equal to the distance to the building minus 150mm.

Fig. 2.6 Example of typical pea gravel used in the bedding of pipework.

Fig. 2.8. Detailing of drainage trenches highlighting minimum distances for drainage situated close to building foundations.

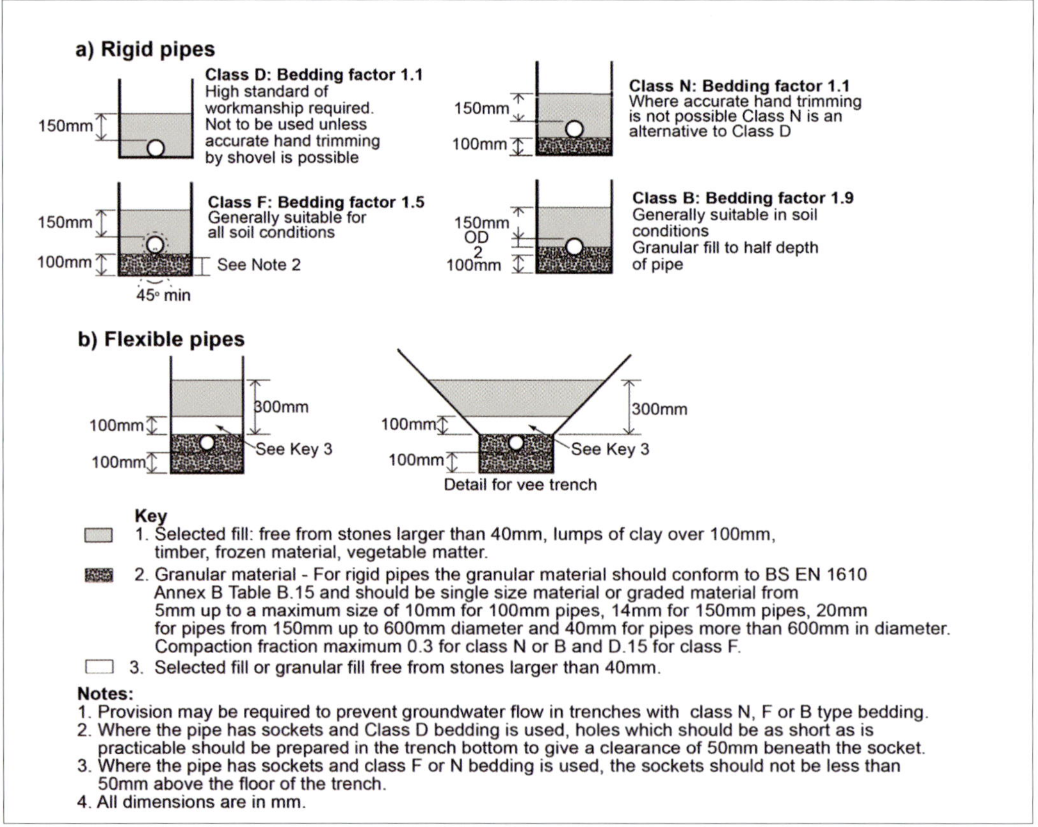

Fig. 2.7 Important considerations when specifying rigid or flexible drainage pipes. (Source: Building Regulations (2010), Approved Documents H)

Pipe Capacity

Drains should be designed to have sufficient capacity to convey the expected flow and volume of waste or foul water. The capacity of the pipe is limited by its diameter and gradient. Criteria that inform pipe capacity are:

Filling degree: The specified proportion of water to air within a horizontal pipe is different in different countries. This is due to the use of different shaped fittings and trap sizes. The ratio of the depth of water to the internal diameter of the pipe is the filling degree. Hence, a pipe full of water has a filling degree of 1 and half-full a filling degree of 0.5. As airflow is normally the limiting factor with any drainage system design, filling degrees for underground pipes are not normally over 0.8. With most drains and sewers, the water levels will vary over time, but combined systems will have the greatest variations due to the nature of rainfall.

Peak flow: This can be determined by the number of dwellings and connected appliances being used at any one time. That is, the number of WCs, baths, washbasins, sinks, washing machines and dishwashers. Examples of flow rates from dwellings with a household group of one bath, one WC, one or two washbasins, one kitchen sink and one washing machine are shown in Table 2.3.

Self-cleansing velocity and deposition: This term 'self-cleansing velocity' is used to indicate the flow rate of water needed to move loose material along a horizontal drainpipe. Most drains or sewers do not run at constant velocities, so such values need to be attained regularly to help clear any deposits that may have accumulated during lower flows. Many deposits are also not simply granular in nature and may be sludge-like or precipitated as layers as the wastewater cools. These may not simply erode away but deposit and reduce the flow depth of the pipe. Hence, in reality most drains are not self-cleansing irrespective of their design flow rates. The latest EN 752 on drain and sewer design now includes the following clause with no numerical value:

Drains and sewers shall be designed to provide sufficient shear stress to limit the build-up of solids to levels which do not significantly increase this risk.

The material of the pipes will initially influence the flow rates, but after a while deposits will build up on all pipe materials. These deposits of detergents, fats and minerals will produce a slime-like covering that will remove any material differences related to friction factors.

Where deposits are anticipated or the volume of water flowing in the pipe is to be minimized, an alternative procedure for cleaning the pipes may be needed. Tipping buckets, syphons or other means of producing significant flushes of water may be a more reliable way of achieving cleansing of pipework.

Pipe size and gradients: The minimum self-cleaning velocity of drains is often given as 0.76 metres per second, and for traps EN 1253 suggests a volume flow rate 0.6 litres per second. Hence it is sensible for pipe sizes and gradients to be selected so that the flow velocity will frequently exceed this minimum threshold. For foul water drains, the velocity should be achieved at least once a day, but for stormwater drains – where granular deposits are most likely – such velocities should be achieved during typical rainfall periods.

The pipe size and gradient are also influenced by the anticipated peak flow. Pipes must be sized to cope with the peak flow of the property or properties being

Dwellings	Number	1	5	10	15	20	25	30
	Flow rate (litres/second)	2.5	3.5	4.1	4.6	5.1	5.4	5.8

(Adapted from Building Regulations (2010) Approved Document H.)

Table 2.3. Example flow rates from number of household groups of dwellings.

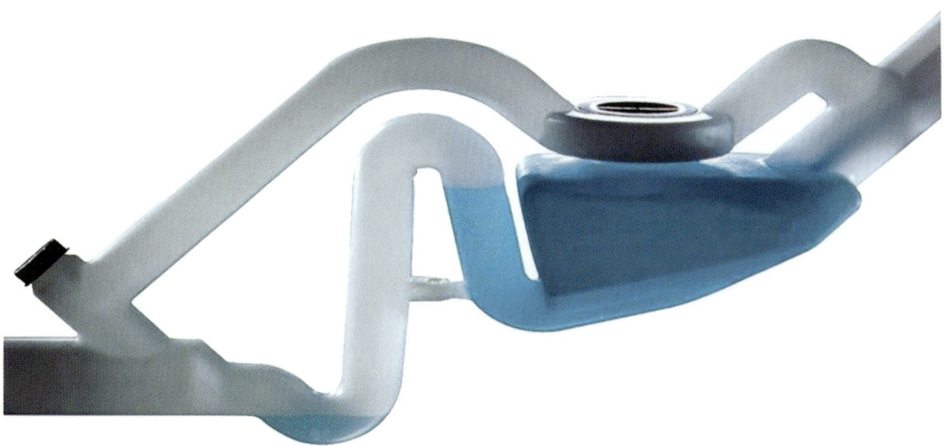

Fig. 2.9 WISA drain syphon that collects 14l of wastewater before quickly discharging the contents along the drain. The device is mounted in a shallow access pit outside of the building. Note how the unit is ventilated.
(Source: WISA sanitar image)

drained. In all gravity systems, the pipework must not decrease in diameter. Normally the smallest diameter will be at the appliance so that all items that pass through the appliance trap should pass freely through the rest of the drain and sewer system. This is why a conventional WC with an outlet of around 80mm will normally be attached to a 100mm discharge pipe. If there are additional WCs and other sanitary appliances, the pipe diameter of the branch or stack may need to be increased.

Some examples are:
- Peak flow < 1 litre/second = 1:40 gradient = 100mm pipe diameter = max capacity 9.2 litres/second
- Peak flow > 1 litre/second = 1:80 gradient (minimum of 1WC) = 100mm pipe diameter = max capacity 6.3 litres/second OR = 1:80 gradient (minimum of 5WCs) = 150mm pipe diameter = max capacity 15.0 litres/second

To adopt the drainage system, historically water authorities have required pipe sizes of at least 225mm to allow for future developments and additions to the system. However, this is no longer the case.

The other consideration is the groundworks cost of deep trenches/gradients versus the cost of shallow trenches. Therefore, in certain instances, a shallow drain with a larger pipe size is specified to minimize costs.

Further guidelines on pipe size and gradients can be found in relevant building codes and standards. For example, England's Building Regulations Approved Document H.

Joints and changes in direction: Joints in drains can be in the form of seals, couplings, clamps, sockets and spigots and anchoring. Change in direction impacts on fluid flow and this should be kept to a minimum. Flexible joints are now more prevalent in sewers to ensure adequate allowances are made for relative movement between the pipes as well as ground movement. Plastic drainage pipes are typically available in 1, 3 and 6m lengths in a range of outside diameters e.g. 82mm, 110mm, 160mm, 200mm, 250mm, 315mm and 400mm. The standard lengths should be considered during design to enable easy installation and assembly. Iron and clay pipes are available in similar diameters and lengths. Today many pipe materials use similar connections, especially push-fit, with a high degree of flexibility that used to be limited to plastics, so changing pipe material is simpler. Modern types of connection include:

- **Push-fit connection.** This means that each pipe typically has a socket end, and a plain spigot end with a captive ring seal that can be pushed together. Although this seems to be a simple system, it can prove difficult in a trench situation as a considerable amount of force is often needed to displace the seal to allow the pipe to enter the fitting. Once the pipe is pushed into the socket it then needs to be retracted a few millimetres to allow for expansion. Such fittings are normally very reliable and if assembled correctly, using a properly chamfered pipe that does not damage the seal, they should be watertight indefinitely.

Originally, push-fit joints used simple 'O' rings, but these tended to need quite high insertion forces to get the pipe to compress the ring. The rings were often supplied separately and needed to have lubricant applied at the time of installation. Also, the force required would sometimes displace the ring as it was inserted, resulting in a joint that could not only leak, but also provide an internal barrier for solids. Today, most wastewater drainage pipes for external installations use captive 'T' rings. These rings are pre-lubricated and are easier to deflect for easier insertion. The degree of chamfering required on the inserted pipe is also now less and the joints are far more reliable. The seals used to be rubber, but are now engineered elastomers that are specifically designed for sealing pipes.

Push-fit pipes are not only limited to plastics pipes, clay and iron pipes now utilize similar fittings that provide a degree of flexibility, and speed of making, that traditional caulked joints did not. Push-fit concrete, clay and iron pipes in trenches can be jointed with the help of mechanical diggers with suitable materials used to protect the ends of the pipes.

Fig. 2.10 (a) Plastic pipes with push-fit joints.

(b) A cast iron stack with a good clamp on fitting.

- **Flanged connection.** The pipes are assembled by screwing, or welding, flanges to the pipe ends. The flanges are then bolted together with the rubber seal/washer providing a complete seal between the flanges to avoid leaks and seepages.
- **Solvent weld connection.** For some plastics such as PVC, solvent cement or welds are used to connect pipes. This method is often favoured by plumbers because the fittings are cheaper and it produces a system that is not readily changed or abused by users. It is a fast wet process that takes skill and requires the pipe ends to be clean and dry prior to connection. With skill and care, solvent weld joints can last indefinitely, but if wrongly made the joint has to be cut out.

The latter two connection methods are typical for rigid pipes. A combination of assembly strategies may be used, especially when working on existing buildings and making connections to existing drains and sewers. It is also worth noting that clay to PVC couplings and adaptors are needed to make connections between different pipe materials. They come in different shapes and sizes and can also be used in instances where it is necessary to join and seal cut pipes.

Flexible, or ridged, pipes may use flexible connections, e.g. like concertinaed hoses, where vibration could be transmitted or when pipes change directions. However, such connections should be minimized as they are more prone to deposition and wear. For changes in directions a wide range of angular, tee, Y, or elbow joints are available. See the section on 'Access for Maintenance' for further requirements where drainage pipes change levels or directions.

Pipe shape: The most common shape for modern pipes is circular. This is the easiest to manufacture, transport and install. However, in a situation with variable flows, as in most drainage systems, it is far from ideal. In Victorian times, sewers were often built as ovoids (egg shaped). This gave a narrow channel at low

Fig. 2.11 Ovoid pipe with narrow width for low flows and increasing widths for most of the height. Note, this pipe is also corrugated as it incorporates multiple inlets at its crown.

flows, but a much wider one at times of heavy flows. The work required to produce such sewers by hand with brick was considerable, but impressive. Today some concrete and plastic pipes are available in ovoid shapes, but they tend to be at least 0.5m high and are intended for stormwater or sewage systems.

Why are drains normally circular? Well, originally drainpipes were made from tree trunks and were hollowed out by hand. Hence, the circular shape as it was the natural shape of the hollowed-out wood.

In some countries square pipes are common. For example, in the Caribbean the islands are subjected to very heavy rainstorms and tropical storms, but generally the air temperature is high (well over 20°C). So, channels to divert these 'flash floods' are formed of concrete troughs with concrete slabs laid on top to provide firm footpaths. As freezing is not a problem, gradients are not important as any residual water will soon evaporate. As long as the channels lead downhill, typically towards the sea, they function as required.

The Use of Grease Separators with Drainage Systems

The problem of fats, oils and grease (FOG) in drainage systems is not new, but changes in cooking methods and detergent innovations over the years have tended to increase the problem. In the past, most washing of clothes would include a 'boil wash' at a high temperature, but the drive towards energy saving has led to the development of low-temperature washing methods at the same time as the number of fast-food outlets has grown rapidly. The increased fat content of cooler wastewater has contributed to the problems of fatbergs in our sewers and the clogging of smaller drains.

Increasing the temperature of the wastewater could reduce the deposition in drains, but would increase energy consumption and encourage more heat recovery systems; so would probably not help. Stopping the grease entering the drains is the better option.

Although very little has changed in Building Regulations requirements over the last few decades, the guidance and recommendations have often changed to keep pace with innovations and the publication of new Standards. Grease traps, or interceptors or separators, are a clear example. Up until 2002 they were not mentioned in the English and Welsh Building Regulations. Under the topic of Layout, the following recommendation was introduced in AD H1:

Drainage serving kitchens in commercial hot food premises should be fitted with a grease separator complying with BS EN 1825 -1: 2004 and designed in accordance with BS EN 1825-2: 2002 or other effective means of grease removal.

This recommendation, and the referenced Standards (BS EN 1825-1:2004 Grease separators. Principles of design, performance and testing, marking and quality control, and BS EN 1825-2:2002 Grease separators. Selection of nominal size, installation, operation and maintenance) are still current in 2019.

An industry has since developed around such grease separators providing services for the recovery of the collected grease for recycling. This helps to facilitate the regular maintenance that such separators need. If the grease is not removed regularly the separators will simply discharge the excess grease-laden wastewater to the sewer, causing more problems downstream.

Grease separators are essentially a large trap that holds a considerable volume of water. Some are installed within the building and others are designed to be fitted outside and buried. Wherever they are located it has to be ensured that the wastewater entering them is warm enough for the FOG material to reach the separators, this may require the use of insulation as well as trace-heating in some cases.

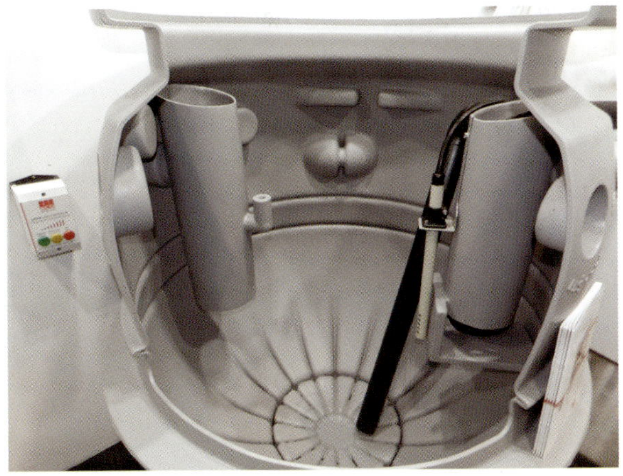

Fig. 2.12 A section view of a grease interceptor that would be installed in the ground. It is basically a settlement chamber fitted with a probe to detect the level of accumulated grease. The chamber needs to be emptied by a tanker and then refilled with clean water.

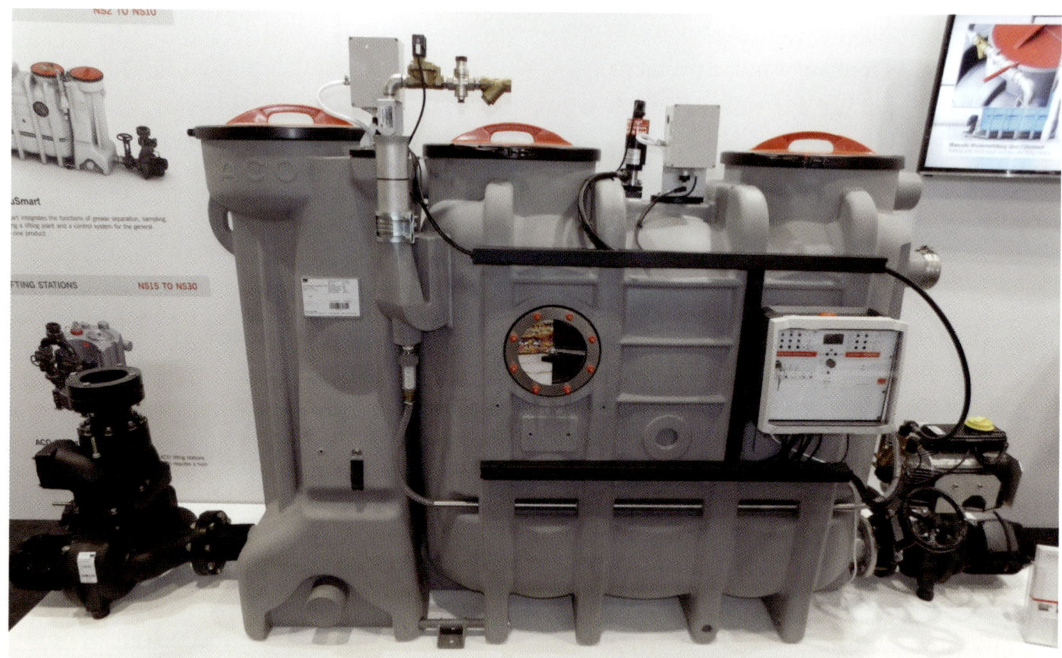

Fig. 2.13 A fully automated grease interceptor that is to be installed inside a building, often in the basement. This unit will monitor the level of accumulated grease and when at capacity it will instigate a safe discharge of the grease using its own pumps. Once the chamber is emptied, a high-pressure jet cleaning process will take place before the chamber is refilled with clean water.

However, grease separators all operate in similar ways. As the new wastewater enters the separator the less dense FOG material should float to the surface and the remaining wastewater should descend to the trapped outlet to drain. The sophistication of these devices varies considerably: from simple settlement chambers to fully automated containers that incorporate various sensors and controls to ensure consistent reliable operation.

Access for Maintenance

It is important that all drainage systems, internal and external to the building, provide adequate access for periodic inspection and maintenance.

According to the Approved Document H, access points should be provided:

- on or near the head of each drain run, and
- at a bend and at a change of gradient, and
- at a change of pipe size (but see below if it is at a junction), and
- at a junction unless each run can be cleared from an access point (some junctions can only be rodded through from one direction)
- Principally, access should be provided to long runs of drains or channels.

In Victorian times, lamp holes were incorporated into some sewerage systems so that a visual inspection could take place from the surface. More recently, rodding eye systems that allow drain rods to be inserted for blockage clearance have become popular. Systems, such as the Marscar Bowl, which uses a semi-spherical chamber at just below ground level and a 45 degree outlet, connecting to a lower-level drainpipe using another 45 degree connection, are sometimes used in hilly locations so that many buildings can be drained easily but the resulting depth of inspection chambers is minimized enabling access to be obtained at ground level.

The access points shown in Fig. 2.14 are typically located in drainage systems external to the building. These can be found in the grounds of the building up to boundary property as well as on public pavements

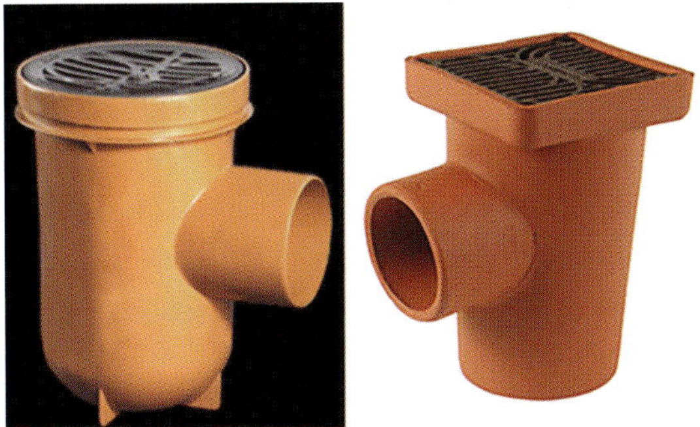

(b) Access, roddable gully.

(c) Hopper: square.

Fig. 2.14 Drainage access components. (a) Access chamber: cover, riser and base.

(d) Rodding eye.

and highways near the property.

Rodding Points

Also known as rodding eyes, these provide an access point with a removable cover and can be found on a simple horizontal, or steeply angled, pipe that branches off a main drainpipe. They are in essence a capped extension of a drainpipe.

(Shallow) Access Chambers and Inspection Chambers

Drainage systems are generally built a metre or so below the ground level and tend to follow the contours of the land. However, where there are significant undulations of land or the drains or sewers are particularly deep, simple shallow gravity systems may not be possible. With below-ground drainage systems, or buried drains and sewers, traditionally the pipes were laid at a normal slope of around 1 in 60 and inspection chamber shafts – often with step irons – would be sunk down to the invert level. This was simple but created hazards for any person having to enter the drainage system for maintenance.

Access chambers provide access for inspection and pipe/drain rods at a shallow level. Comparatively, inspection chambers have more branches feeding them so they are deeper and larger. Inspection chambers are deeper (1.5m +) and wider than access chambers. They are typically constructed with brick, pre-formed plastic or precast concrete. They provide access for maintenance and are big enough for service engineers to enter to carry out work on drains. They are commonly used in public sewers, i.e. beyond the property boundary where drain channels are benched and specifically bedded in cement/sand material with a slope so that spillage can flow back into the drain. There are also backdrop inspection chambers that are situated where there is a change in level in drains, i.e. a branch drain is at a higher level to a main drain. These are available in two types, internal and external. External backdrops tend to be new-build installations where the downpipe can be located outside the chamber. Retrofitted pipework is more common in internal backdrop chambers as less excavation is required.

Although, traditionally, inspection chambers were used for access for maintenance of drains and sewers, recent health and safety legislation around working at heights and in confined spaces has made the provision of person entry chambers something to avoid. Hence, shallow chambers – that can be accessed from the surface without the need to enter – and rodding eyes are now more popular. Where a person is required to enter a chamber, breathing apparatus and safety harnesses may be required.

The minimum dimensions for access fittings and inspection chambers are specified in England's Approved Document H (Table 2.4).

Once laid, all drains are tested for flow and watertightness. This can be done using air, water/dye or smoke test. Although, modern techniques also include manual or automated CCTV surveys.

Fig. 2.15 (a) A conventional inspection chamber, constructed from pre-cast concrete components. These include the base, rings, biscuit, frame and cover.

(b) A galvanized steel, sealed cover that is intended for non trafficked areas.

(c) Fitting a resilient seal to an inspection chamber base, an alternative would be to use a mix of fine cement, especially if the side of the chamber is to be cut for the insertion of an additional pipe. (Source: Marshals)

Type		Depth to invert from cover level (m)	Internal sizes		Cover sizes	
			Length × width (mm × mm)	Circular (mm)	Length × width (mm × mm)	Circular (mm)
Rodding eye			As drain but min. 100			Same size as pipework[1]
Access fitting						
Small	150 diam.	0.6 or less				
	150 × 100	except where	150 × 100	150	150 × 100[1]	Same sizing as access fitting
Large	225 × 100	situated in a chamber	225 × 100	225	225 × 100[1]	
Inspection chamber						
	Shallow	0.6 or less	225 × 100	190[2]	-	190[1]
		1.2 or less	450 × 450	450	Min. 430 × 430	430
	Deep	> 1.2	450 × 450	450	Min. 300 × 300[3]	Access restricted to max. 350[3]

Notes:
1. The clear opening may be reduced by 20mm in order to provide proper support for the cover and frame.
2. Drain up to 150mm.
3. A larger clear opening cover may be used in conjunction with a restricted access. The size is restricted for health and safety reasons to deter entry.

Table 2.4. Minimum dimensions for access fittings and inspection chambers.
(Source: Table 11. Building Regulations (2010), Approved Documents H Drainage and waste disposal)

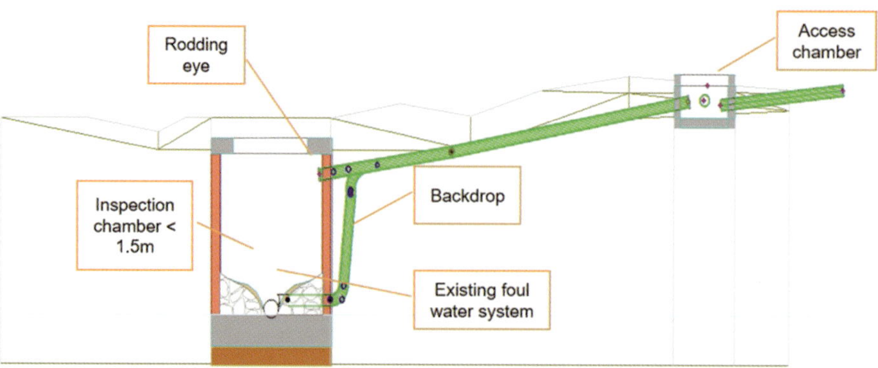

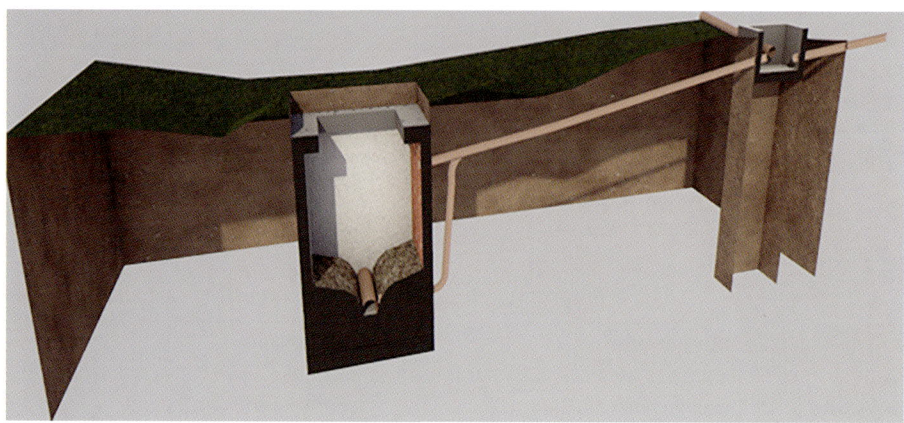

Fig. 2.16a-b Inspection chamber with external backdrop and access chamber.

Chapter Summary

This chapter has covered the engineering aspects of drainage installation. The various types of pipe and connections have been considered. It has shown that pipework has developed over the years and today plastics pipes and push-fit joints are now established. This has affected the methods used to protect pipes and buildings when burying drainage pipework.

The hydraulic design of systems is based upon fluid dynamic principles and rules of thumb that have been developed over the years. As wastewater systems deal with the multiphase flow conditions of air, water and various forms of waste, much empirical work has been used to simplify or enhance drainage design. Standard-ization work around the world has brought together many national traditions and practices to create a more global approach to drainage design. Although computer modelling has developed greatly in recent years, pipe size availability has limited such methods to large-scale projects and academic research.

The issue of access to drainage systems for inspection and maintenance has evolved. Due to significant changes in health and safety culture and legislation, along with the increase in availability of inexpensive CCTV systems, man-entry chambers are no longer the norm. Surface access provision, especially for domestic drainage, is now normal practice. Where access by

Fig. 2.17 Internal backdrop manhole with discharge onto the benching and an access fitting at the top is shown in the left-hand image. The right-hand image shows evidence of an external backdrop, with the rodding point near the top and the outlet near the benching.

persons is still needed there are requirements for extensive safety equipment. This may include personal protective equipment such as breathing apparatus, body harness, winches and gloves.

SURFACE WATER DRAINAGE

The external building site can be left as:

- Hard standing, e.g. pavements
- Soft standing, e.g. lawns
- Or a combination of the two.

The purpose of surface water drainage is to provide temporary storage to reduce the impact of any rainfall on the surrounding areas. If the water level at any point in a surface water system is observed over a period of time, a graph can be plotted of the water level with the rainfall over the same time superimposed. Such a diagram is known as a 'hydrograph'.

The hydrograph of any stormwater event can be modified beneficially by the use of: green and blue roofs, temporary storage containers (ranging from water butts, crate structure, and garden ponds to floodable sports pitches, swales and infiltration beds), hydro brakes, wetlands, etc. Such hydrograph modification tools can be applied to both new-build and existing areas of construction.

Surface water drainage should no longer be considered in isolation from other environmental issues. Rainwater needs to be considered as an important and useful resource, not a waste product that simply needs to be collected for disposal. Increasingly, surface water management is being considered under flood management or whole river basin and water catchment management schemes. Similarly, rainwater harvesting systems are more likely to be considered for a large area scheme than an individual dwelling due to economies of scale and central maintenance. In existing areas, downpipe disconnections are being used as a simple way of reducing the volume of water going directly to the drainage system or soil. The disconnected downpipe can be used to feed a rain garden, an ornamental rainwater pond, a flowerbed or used to permanently irrigate crops. In highway management, rain gardens are now being used to help clean run-off from roads and provide improved visual amenities and habitats for small birds and insects. Therefore, collecting run-off in a subsurface storage structure helps reduce flood risk and provides a source of water for tasks such as vehicle washing and WC flushing.

Surface water drainage is briefly covered in EN 752-4:1997 (Drain and sewer systems outside buildings – Part 4: Hydraulic design and environmental considerations). The useful standard on issues of prevention of flooding from a number of sources is the BS 8533:2017 Code of practice on assessing and managing flood risk in development.

Sustainable Drainage Systems (SuDS)

Sustainable Drainage Systems (SuDS) were originally known as Sustainable Urban Drainage Systems (SUDS), but they are no longer limited to urban situations. Today, SuDS have now been expanded to Water Sensitive Urban Design (WSUD) and Human Sensitive Urban Design (HSUD) systems in which a more inclusive approach is taken with the water supply and drainage, which is central to the quality of life of people as well as the quality of the built and natural environments.

The main aim of any type of sustainable drainage system is to mimic the natural processes that deal with water in undeveloped areas and incorporate them into habitable areas in a sympathetic and effective way. Whereas most traditional stormwater engineering systems use pipes and concrete structures typically buried and hidden underground, sustainable systems rely more on softer engineering utilizing the principles of infiltration, retardation and dispersion of water.

The underlying principles of SuDS is to deal with surface water as close as possible to its source and not simply passing it on to another area for treatment or disposal. In instances where it cannot be dealt with at source, it should at least be decreased in volume and contamination before discharge. The concept of SuDS is based on the combination of three objectives that need to be balanced:

- Water quality
- Water quantity
- Biodiversity and amenity

The rise in importance in SuDS has coincided with increased awareness of issues with excessive surface water, such as flooding. In Europe, the Water Framework Directive has focused attention on river basins and catchment areas, so SuDS are now becoming frequently integrated into local planning and master planning designs. The early integration of SuDS into any development is important, as retrofitting SuDS is very difficult, disruptive and expensive. Some regeneration schemes have included urban greening of existing culvert or canalized surface water features. In addition, SuDS features such as rain gardens have been incorporated into exiting streets to improve the environment, introduce some traffic calming and help improve air quality.

One of the main obstacles to the proliferation of SuDS has been the issue of who looks after the system after it is installed. Traditionally, pipework installed by a builder is connected to a sewer and the local wastewater company will look after the pipes as long as they meet various criteria. The criteria the wastewater companies' use is set out, in the UK, in Sewers for Adoption. Edition 8 of these guidelines now includes criteria for the adoption of certain SuDS elements, but not all.

SuDS Train

In SuDS, there are two basic trains; the management or control train and the treatment train.

The control train has three elements in sequence:

- Source
- Site
- Region

Between each element is a conveyance system. Table 3.1 gives examples of the elements that might make up or contribute to these items.

The treatment train is similar and comprises:

- Collection
- First treatment
- Second treatment
- Third treatment

However, not all the elements may be required in all circumstances and some can be combined. The elements are also those in the management train.

How much and by what process the treatment is

Management train element	Typical feature	Comment
Source	Green roof, blue roof, permeable paving.	Water reuse or harvesting can also minimize the volume to be passed to the next element.
Conveyance	Natural water courses.	Pipes may also have to be used when directing flows below highways or foot paths as well as through buildings, etc.
Site	Ponds, basins.	
Conveyance	Natural water courses, swales.	Brings together various sites.
Region	Wetlands.	

Table 3.1 Management SuDS train.

Treatment train element	Typical feature	Comment
Collection and first treatment	Green roof, blue roof, permeable paving.	Filtration in these features will remove large debris, such as leaves and provide limited removal of contaminants.
Conveyance	Natural water courses.	Travelling over or through ground will help remove more contaminants. Where natural water courses are used dilution will also contribute.
Second treatment	Ponds, basins.	Travelling over or through ground will help remove more contaminants. Where water-filled ponds are used dilution and natural aeration will also contribute.
Conveyance	Natural water courses, swales.	Travelling over or through ground will help remove more contaminants. Where natural water courses are used dilution will also contribute.
Third treatment	Wetlands, ponds.	The long residence time provides final treatment and/or polishing before discharge.

Table 3.2 Treatment SuDS train.

made depends on the level of initial contamination and the sensitivity of the receiving water. Rainwater from a roof will require little treatment, but run-off from a contaminated area, such as a fuel station or refuse site, may require removal of hydrocarbons, chemical and heavy metals.

Design Considerations

Surface water drainage is often barely visible. However, the structures beneath the surface can be considerable. Rainwater that falls on hard surface areas, such as pavements and roads, enters the underground drain via trapped or untrapped gullies. Channels with a slight fall are often provided between gullies to encourage flow and avoid 'puddling'. Special gullies known as petrol interceptors are sometimes installed as additional precautions are required in car parking areas. This is to prevent petrochemical contaminants entering into the sewerage or infiltration system.

The sustainability of any drainage system depends on the materials used in the pipework, backfilling and treatment. Life cycle analysis can show the impact of materials on the environment and people in the areas where the raw materials are obtained, along with the impacts of transportation, installation, use, demolition and final disposal. Sustainability is not limited to minimizing water use.

The main design considerations for SUDs design are: rainfall intensity, run-off and infiltration.

Rainfall Intensity

As with roof drainage design, rainfall intensity is important. However, for surface water, the units used are different as, in general, the areas involved are much larger. Whereas for roofs the rainfall intensity (r) is normally given in litres per second per square metre, for surface run-off the rainfall intensity (i) is given in litres per second per hectare. [1 hectare = 10,000 m2]. The equation used to determine the peak flow from an area of less than 200ha is:

$$Q = \Psi i A$$

Where Q = peak flow rate (l/s)

Ψ = run-off coefficient [between 0 and 1]

i = rainfall intensity (l/s/ha)

A = horizontal area receiving rainfall (ha)

Fig. 3.1 The linear drainage towards the gulley appears to be a simple slot.

Fig. 3.2 These linear drainpipes show that beneath a simple slot can be connections to either a round or ovoid pipe of various sizes. Depending on the use of the surface, the pipe may be simply bedded or enclosed in

Description of area	Run-off coefficient ψ
Flat roof < 100m2	1.0
Large flat roof > 10,000m2	0.5
Impermeable areas	0.9-1
Permeable areas	0-0.3

Table 3.3 Run-off coefficient values.

Some examples of run-off coefficient values are in Table 3.3.

For areas of more than 200 hectares, flow simulation modelling by expert consultants is suggested. For most sites, there are two options for managing surface water; drain or infiltration. Although traditionally the first choice was to direct it to a nearby drain or sewer, modern practice is to try to deal with the water as close to the source as possible. Hence, infiltration to soakaways is now a preferred option, especially if it can be part of a SuDS scheme.

Run-off

Run-off rates are also key to an SuD system. The run-off due to rainfall will be dependent on the amount of rain that falls. In designing any rainwater system, a limit has to be imposed, so there will always be some rainfall events that will overflow a system. If a system were designed to accommodate all possible rainfall it would be so large that it would not work well at normal or low rainfalls and be very expensive. The added complication today is that climate change is altering the volume of water that can fall at any one time (although the distribution of rainfall is changing, the average annual rainfall for an area remains consistent over many decades). Hence, although storms with various return periods can be considered, 1 in 100 years is a typical choice. However, 1 in 500 may now be a more appropriate design base.

Once the rain has fallen, it will run off the land at a rate that depends on the surface. The baseline for run-off from a site is the Greenfield Run-off Rate (GF). This is how fast water will run off a site in its natural state, prior to any development. The units are litres per second per hectare. Once a site is developed, it will have an Urban Run-off Rate (U) of about twenty to thirty times the GF.

If a site has previously been developed, the aim should be to reduce its run-off to the GF, but if that is not possible a 'betterment rate' should be achieved. This is the reduction factor in the peak run-off rate of the existing site. The peak rainfall intensity can be assumed to be 75mm/hour.

To reduce the run-off from the site, a number of devices can be used to slow the flow. These include:

Weirs: Usually watercourse-wide, barriers or walls over which water flows. Some weirs can include adjustable elements to change their height and control the flow.

Fig. 3.3 Example of different height weirs used to provide sequential flow through a structure.

Perforated risers: Similar to weirs, but with holes in their faces to allow water to not only flow over them, but also through them. In periods of low flow they act as a course filter, but in high flows they produce waterfall effects like a weir.

Orifice plates: Barriers across the watercourse with a set, or adjustable, width opening. In SuDS they tend to be square in shape and not circular as found in supply pipework systems.

Hydro brakes: Complex funnels that entrain air to produce a vortex or vortices that may be specified or adjusted to specific discharge rates. They tend to be used at outlets from retention ponds or at inlets to water courses. They may be fitted with overflows, or emergency access doors, so that in the event of a blockage water can be drained from the site.

Whatever mechanisms are used, the minimum final discharge from a site should be 5l/s. A SuD system may need a number of flow controls, incrementally increasing in flow rate capacity as they progress down each stage of the system's management train. At the beginning of the MT (source control components) flow rates are likely to be less than 5 l/s.

Infiltration

The infiltration principle is well accepted in stormwater disposal. Rainfall naturally evaporates and infiltrates at various rates depending on the humidity, temperature, soil permeability and the level of saturation. As water enters the soil it will percolate in various directions, depending on the soil structure, soil type and level of flooding.

Within SuDS, there are many features that use infiltration as a main or secondary function. These include:

Swales: Grassed or vegetated channels with a flat base that can collect, treat, store and convey water. The maximum linear gradient should be around 1:50 and side slopes of 1:3 for safe access. For effective treat-

Figs 3.4a-b Hydrobrake installation, b. An in situ shot of a water attenuation valve complete with its overflow pipe and servicing handle. The attenuation device is fitted in a chamber, for access, and the outlet is behind the unit. Water enters at the side normally and as the level of water increases the water flow creates a vortex that impedes the flow. If the unit blocks, the door at the front of the unit can be opened to gain access to the unit. If more involved work is needed to remove an obstruction, the whole attenuation device can be lifted out using the rod on the back wall of the chamber.

Fig. 3.5 On-site water-sensitive landscaping showing the mix of natural and recycled landscaping.

ment, the minimum retention time should be ten minutes.

Under-drained swales: Grassed or vegetated channels with a flat base that can collect, treat, store and convey water with a filter drain below the surface that receives its water by infiltration.

Filter strips: Grass or vegetated verges. Normally located along the edge of hard surfaces that allow water to flow as a sheet to another SuDS feature, such as a swale or filter drain. The drop from the hard surface to the start of the filter strip should be at least 20mm. The strips will also infiltrate to the ground and provide some treatment as the water flows through the vegetation. The maximum linear gradient should be around 1:50 and side slopes of 1:3 for safe access. For effective treatment, the minimum retention time should be ten minutes.

Filter drains or 'French' drains: Trenches filled with

Fig. 3.6a–b Swale and a pre-formed swale inlet component.

Fig. 3.7 Example of a filter drain at the edge of a roadway.

Fig. 3.8a Demonstration permeable pathway showing the construction layers.

Fig. 3.8b Section through a permeable pavement, showing the various layers that lie beneath the visible blocks.

open-graded stones that allow run-off to flow into them and either: infiltrate directly into the ground, or travel along the drain to an outfall. To protect the stones from excessive silt, a top layer of sacrificial stones over a geotextile may be used and replaced as required without replacing the whole system.

Permeable pavements: These comprise a series of construction layers to intercept, treat and store rainwater. They are often used for driveways, car parks and other lightly trafficked areas.

Infiltration and retention basins: Open grass depressions, which are normally dry. However, during periods of heavy rainfall they can provide temporary storage for floodwater, before slowly allowing water to dissipate into the ground (if an infiltration basin), or, if they are part of a conveyance system, release it into

the next stage of the drainage system via an outlet (if a retention basin). The typical gradient of such basins is around 1:3.

Infiltration systems do not have to be large and visible. Infiltration inspection chambers are now available that can be installed below roadways and slowly dis-

Fig. 3.9a An infiltration basin in the USA.

Fig. 3.9b Small infiltration basin with hydrobreak outlet in distance and pre-formed swale inlets to the right.

Fig. 3.10 A London rain garden showing the bump-out gap between the pavement kerb and the rain garden kerb that allows water to enter the garden. The garden kerb is angled to create accessible parking places between the rain gardens.

Fig. 3.11 Not all shrubs and plants are suitable for urban rain gardens. This one has overgrown its garden reducing parking spaces and the roadway. Rain gardens in such locations need to be tended regularly so that they enhance the area and do not become eyesores.

charge any water collected in the kerb gullies. Also, retention tanks can be constructed beneath garages and other structures, which temporally hold rainwater from the building before slowly discharging it via a hydro brake to a sewer or water body, or to an infiltration system.

Bio retention: A structure that is generally used in urban areas to collect and treat rainwater, using landscape features to perform a drainage function. The construction usually has a vegetated layer, often containing trees, over a free-draining soil. Water infiltrates through the soil in the root zone into the underlying drainage layer before discharging into the next stage of the drainage system.

Rain gardens: An area where rainwater is collected and infiltrated, but supports a range of plants that enhance the location. An overflow should take any excess water to the next stage in the drainage system.

Ponds and wetlands: Bodies of open water designed to accommodate water during periods of heavy of rainfall. They also provide mechanisms for the treatment of water and the removal of silts and pollution, as well as infiltration when designed to be permeable. Their design should also enhance visual amenity and provide biodiversity. If the cover of vegetation exceeds 75 per cent a pond can be classed as a wetland. Ponds should be between 400–600mm deep. Wetlands tend to be about 300mm shallower than ponds; at around 150mm. To provide treatment a reasonable retention time and long flow path should be used. With a wet-

Fig. 3.13 Underground rainwater infiltration tank, only showing the access cover.

Figs 3.12a-b. Examples of ponds.

land a minimum length to width ratio of 4:1 is suggested.

Soakaways: As discussed in the previous chapter, a soakaway is typically a hole in the ground in which rain and surface water is channelled. They are made of porous material and/or filled with gravel or granular soil. The collected water is then slowly released back into the surrounding ground. These are a compact form of Sustainable Drainage System to deal with stormwater at source. Soakaways are not suitable for clay or waterlogged soils or areas where the water table is high.

In the UK, one of the main guidance documents is BRE's Digest 365. In 2016, it was revised and now includes SuDS and flood management aspects as well as climate change. As soakaways are vital to stormwater disposal, stormwater management and environmental protection their design, installation and maintenance needs to be considered carefully.

'Crates' or geocellular structures: Open structure modular blocks, usually made of plastics, that were originally only used for stormwater attenuation, but after over forty years of use in Europe such items have become building blocks for various types of water management system. The open nature of the blocks makes them very light and easy to handle. Most manufacturers make the units stackable so that almost any shape of structure can be produced. To prevent clogging and protect the units from rough backfill, most assembled units are wrapped in various types of geotextile. Within SuDS, various sized and shaped units are often used beneath swales, permeable paving, green roofs and buildings to serve different purposes.

If they are wrapped in a permeable geotextile, they are often used as attenuation devices for infiltration. If they are also topped with layers of treatment media, to remove heavy metals or other specified pollutants, the system can improve water quality prior to infiltration.

If wrapped, fully or partially, in a watertight membrane, they can be used for water storage and often used as part of a rainwater reuse system. If water is directed to them by pipes, they can be beneath impermeable surfaces, such as roads and buildings and may also incorporate channels for various utility services. In areas where trees are planted, the units may be linked with a root watering system and under green roofs the units can be fitted with wicks that can draw water into the root structure of the planted vegetation.

All structures that use full sized units are normally made accessible for maintenance. This may be manually or increasingly using robotic cameras. As the units are below ground, UV degradation is avoided and the underground environment should contribute to a very long lifetime for such structures.

Maintenance

With all these features, different maintenance activities will be required. Typical activities include:

- Litter and debris removal
- Grass cutting
- Wildlife area trimming
- Inspection and maintenance of all inlets and outlets
- Sweeping of hard surfaces
- Leaf collection in autumn
- Removal of silt from silt traps
- Wetland vegetation cutting.

The frequency of these tasks will vary according to the design of the system, location and time of year.

Chapter Summary

Surface water drainage is now dominated by sustainable drainage systems (SuDS). In the past, surface water was sent to surface water sewers and discharged to water bodies. This resulted in flooding during heavy rainfall and little utilization of the water. To control the surface water and reduce its impact on buildings and people through flooding, the SuDS approach has been adopted. This aims to treat and contain the surface water as close to its source as possible and dispose of any excess water in a prolonged duration to minimize the impact on downstream structures and environments.

The various components of SuDS have been examined and brief descriptions given. Some of the issues with SuDS, such as adoption and maintenance have been addressed.

Fig. 3.14 An assembly of stormwater 'crates', showing that when constructed they are capable of supporting traffic.

Chapter 4

RAINWATER DRAINAGE

Roof drainage design involves the appropriate selection and specification of structure and materials. Effective design is an important safety issue because poor drainage design can lead to problems, such as:

- Roof defect and collapse, especially flat ones
- Water and damp ingress in walls
- Overflowing of gutters
- Flooding of buildings.

Guttering System in Houses

Rainwater drainage is provided by an assembly of gutters and downpipes. Rainwater falling on pitched and flat roofs is typically collected by an assembly of gutters that discharge through downpipes to drains, sewers or soakaways. In Europe, the principles of design are set out in EN 12056-3 Roof drainage, layout and calculation. This standard requires the design flow rate for a roof to be appropriately defined and specified. Once the roof design is complete, the main design decisions include how the rainwater will be captured and channelled to the appropriate drainage system. The mechanism with which rainwater is managed in buildings is simply referred to as guttering.

Guttering can take a number of forms, but, arguably, most low-rise buildings are drained using eaves gutters along the lower edge of a sloped roof. Other popular forms of guttering are:

- Valley gutters – between two sloping surfaces
- Parapet gutters – along the lower edge of a flat roof-but within a retaining wall so that the gutter discharges from a weir or outlet spout.

The design and materials for the guttering system depends on:

- Local weather, e.g. the amount of rainfall
- Type of roof, e.g. roof pitch, covering material
- Roof shape, size and area
- Carrying capacity required, i.e. the ability to handle the rainwater run-off.

Different guttering solutions exist to suit the roof type: pitched, flat curved, function (e.g. intensive and extensive green or blue roofs), aesthetics (e.g. visible or hidden). Gutters for industrial buildings are generally larger versions of domestic gutters, but they may use very different forms of outlet, such as a syphonic outlet.

Roof Drainage Components

Roof drainage components consist of guttering and downpipes, gutter outlets and fittings and gullies. They are broadly classified as the gutter assembly (gutters, outlets and fittings) and the downpipe assembly

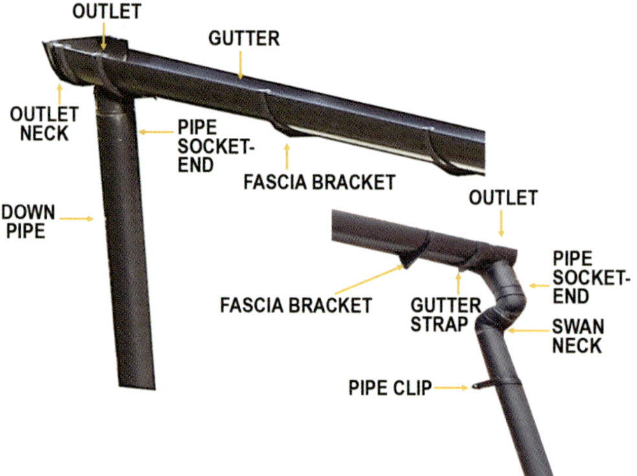

Fig. 4.1a Roof drainage components.

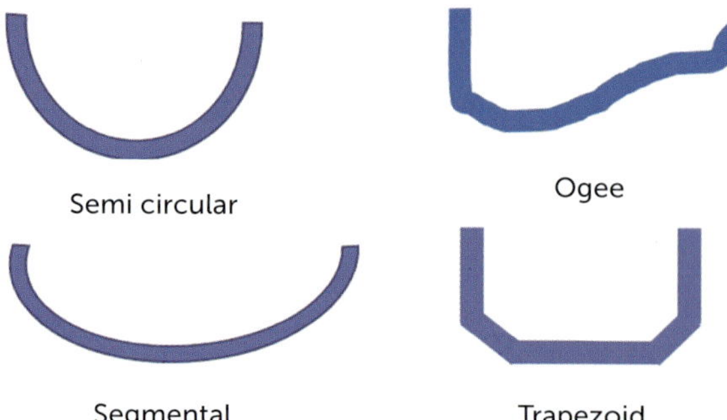

Fig. 4.2 Roof drainage gutter profiles.

(downpipes, gulley and fittings).

Guttering

Guttering materials are historically made from a variety of materials including clay, wood, lead, fibreglass and various metals, e.g. aluminium, galvanized steel, coated steel, copper, cast or spun iron and lead. The material is often selected due to:

- its aesthetics
- how it will weather (perhaps change colour)
- its strength in areas liable to impact
- presence and absence of joints (for extruded gutters made on site from steel or aluminium spools of metal)
- compatibility with an existing system
- Technical quality and degree of required workmanship.

However, domestic guttering is now typically made of plastic as this is cheap, lightweight, easy to fix, durable, low maintenance, and can be formed into different colours. The use of plastic also means that guttering now comes in different shapes or profiles to suit the aesthetic needs of the building as well as provide adequate bearing capacity relative to the roof design.

Gutters are generally classified as semicircular or trapezoidal in cross-section, but there are wide variations due to styles. The popular ogee form is a combination of semicircular and trapezoidal. Gutters are not always visible from the ground; eaves gutters and parapet gutters can be difficult to differentiate in some situations.

Outlets

The two basic types of traditional gravity outlets have round or squared corners (Fig. 4.3).

The round-cornered outlets are more commonly used as the square ones are less hydraulically efficient and tend to impede flow. The rounded edges are preferred to ensure better flow into the downpipes. However, as most outlets are rarely needed to run full, the selection of outlet is mainly due to aesthetic preferences. Conical outlets are also available and the profile of the outlet can have a major impact on its flow capacity. Most manufacturers will provide flow capacity data on their gutters and outlets. There are also other outlet fittings to accommodate angles and stop-ends in the guttering.

Different gutter spigots are connected using socket

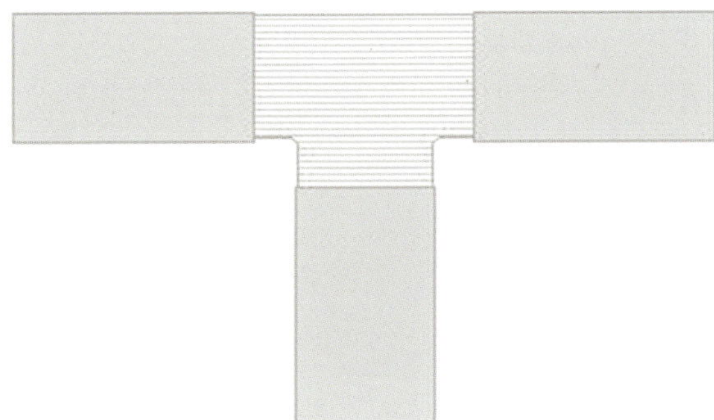

Fig. 4.3a Square-connected gutter outlet.

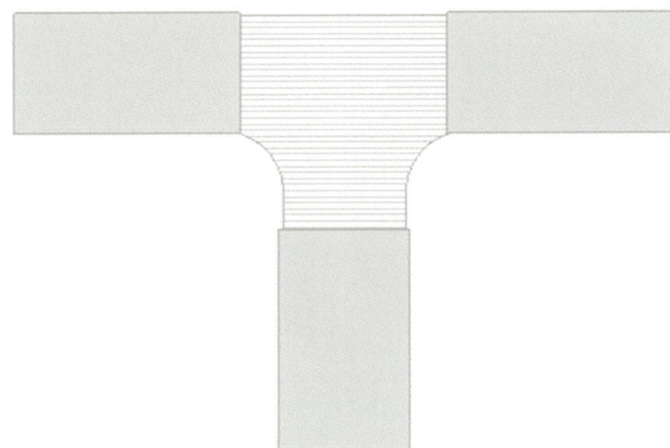

Fig. 4.3b Chamfer-cornered gutter outlet.

ends and straps, or they are bolted together or sealed with a mastic. Gutters made of plastics need to have provision made for expansion otherwise they will generate stress on the fixings and noise during times of expansion or contraction. Expansion joints normally comprise a lubricated resilient seal strip in the socket, with the outside of the gutter held by clips within the socket but with a gap to accommodate the thermal movement. Plastic gutters are prone to thermal movement due to the tracking of the sun, especially in unshaded south-facing situations. Although metal products will also experience thermal dimension changes, these are far less of an issue. Fascia brackets are needed to connect the gutters to the fascia boards. The spacing of the brackets depends mainly on the material of the gutter as well as its length or size. Lighter materials may require fewer brackets, but if they lack rigidity additional fixings may be needed.

Downpipes

Downpipes are located to maximize the flow of water into them from both sides. They should ideally be equally spaced along the gutter but away from the ends.

Swan necks (Fig. 4.4) moulded as one piece, or formed of several pieces are used to accommodate eave overhangs. However, they may become blocked by roof junk and debris. Therefore, to avoid blockages wide-angled connections should be designed and installed.

Socket ends of pipes are connected with pipe clips and screwed to the wall.

Joints on external downpipes to a building are not normally sealed. This serves as a visual warning if a blockage occurs below any joint and makes removing pipes, to clear the blockage, easier. However, if they run internally or in-between walls, sockets with ring seals should be used to prevent water damage in the event of a blockage.

Downpipes can be interrupted by using a simple di-

Figs 4.4a–b Swan neck with pipe clips on down pipe.

Figs 4.5a–b Arched swan neck gutters used to complement the architecture, but with very sharp bends that
should ideally be avoided.

verter fitting – for instance, by installing integrated SuDS, rainwater harvesting systems or even a simple water butt.

In the case of water butts, where a separate overflow is fitted to one, the diverter pipe may be laid at a downward gradient to the butt. It is also vital that this pipe is fitted correctly for the system to work. The diversion pipe must be horizontal to enable both water to flow into the butt from the downpipe and out of the butt, when it is full, into the downpipe.

More details on rainwater harvesting can be found in 'EN 16941-1:2018 On-site non-potable water systems. Systems for the use of rainwater', and more details on SuDS in 'The SuDS Manual (C753) 2015', currently available for free download at www.ciria.org/Memberships/The_SuDs_Manual_C753_Chapters.aspx.

Other forms of downpipe interception or disconnection include rain gardens, natural ponds, or window boxes.

Also, downpipes do not actually have to be pipes. In some examples (Fig. 4.7), rain chains that use the surface tension effect, can be used to produce functional and decorative means to bring rainwater down to ground level. At the gutter outlet, a frame is positioned to support the chain and direct the water onto the chain or chains. The chains are also secured at their base and the collected rainwater directed away as required. Such chains are useful in areas where damage due to vehicles or vandals is likely, but can also be used in other places for decoration. The use of a chain system reduces the likelihood of any downpipe blockage and hence can reduce maintenance requirements and costs.

In addition to rain chains, other alternative forms of downpipes can be used (Fig. 4.8). These can be highly decorative as well as functional. The whole downpipe does not have to be replaced; partial replacement can prove to be equally dramatic.

Fig. 4.6a Simple rainwater harvesting by fitting a diverter to a water butt.

Fig. 4.6b Pipework inside a rainwater harvesting tank. Note the calmed inlet at the bottom, the trapped overflow after the leaf filter with access top.

Figs 4.7a-b Rain chains instead of downpipes.

Figs 4.7c-d Types of upper chain attachment.

Figs 4.7e–g Various lower rain chain attachments.

Figs 4.8a–b Heavily sculptured downpipe example.

Fig. 4.9 Open gully with grating.

Gullies

Gullies are used to collect and channel rain and surface water into rainwater drains, soakaways or other infiltration devices. Grating/grids should also be specified to avoid drain blockages. Note that a trapped gully is necessary at the base of downpipes when connecting to combined sewers.

Roof Drainage Design for Pitched Roofs

Rainwater flowing off a pitched roof is usually collected at the eaves in a gutter fixed to the fascia. The gutter is normally laid flat, for maximum capacity, but if a gradient is used it should be laid in two directions towards the downpipe (Fig. 4.10), typically sized 125mm or 150mm. The downpipe distances should also be appropriately spaced to the length of the building. This also reduces the amount of water the lower part of the gutter has to carry and keeps the gap between the gutter and the tiles to a minimum. Gutters are laid to avoid large gaps between the gutter and the tiles as some of the rainwater may not be collected. This is a particular problem in high winds.

Guttering design for a pitched roof is determined by considering:

1. Rainfall intensity

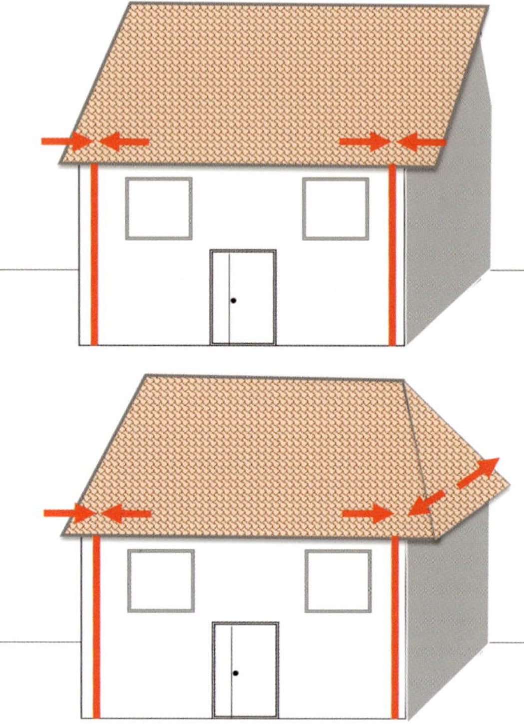

Figs 4.10a-b Examples of water flows in gutters on buildings with pitched roofs and downpipes located away from the ends of the gutters.

Fig. 4.10c. Rainwater drainage layout showing connections to downpipes.

a. The highest amount of rainfall that may fall on the roof

b. Wind-driven rain, particularly on the slope facing the wind (less of an issue in flat roofs)

2. The effective area of roof to be drained

3. The rate at which water leaves the roof.

1. Rainfall Intensity

A simple method for gutter and downpipe sizing is included in the English Building Regulations' Approved Document H.

The simple method uses data from the UK Met Office for low-intensity rainfall. The map used is identical to the NB1 map from BS EN 12056-3:2000 titled 'Rainfall intensity, r, for a 2 minute duration storm event with a return period of 1 year'. So, it should only be used to provide the minimum speciation for an eaves gutter.

1. Calculate the effective area of the roof from the roof plan and multiply it by a 'pitch' factor shown in the Table 4.1a.

2. For designs in the UK, the resulting area is then checked against the Table 4.1b to determine the gutter size (diameter). This should then be correlated to the outlet size (diameter) that discharges to the downpipes.

Other more detailed methods may also be used to calculate the rainfall intensity value, r, depending on the availability of appropriate statistical rainfall data.

If rainfall data is not available

If this is the case a minimum rainfall intensity of between 0.01 and 0.06 l/(s.m2) is chosen, according to local regulations or practice, and multiplied by a risk factor of between 1 and 3, depending on the situation. Using Table 4.2, the equation for roof run-off is:

$Q = r.A.C.$ *risk factor*

Where:

Q = *Flow rate, in litres per second*

r = *rainfall intensity, in litres per second per square metre*

A = *Effective roof area, in square metres*

C = *Run-off coefficient (taken as 1 unless regulations state otherwise)*

If rainfall data is available

Rainfall data should be used when available, but it needs to be selected in line with the appropriate design rate categories, which are similar to the risk factors enumerated above. In the UK, rainfall intensity maps that are produced by Met Office data and as published in BS EN 12056-3 National Annex NB are typically used. It is worth noting that such data is historical and that since the standard was published in 2000 there

Table 4.1a Calculation of drained area.

Type of surface	Effective design area
1 Flat roof	plan area of relevant portion
2 Pitched roof at 30°	plan area of portion x 1.29
Pitched roof at 45°	plan area of portion x 1.50
Pitched roof at 60°	plan area of portion x 1.87
3 Pitched roof over 70° or any wall	elevational area x 0.5

Table 4.1b Gutter sizes and outlet sizes.

Max effective roof area (m²)	Gutter size (mm dia)	Outlet size	Flow Capacity (litres/sec)
6.0	-	-	-
18.0	75	50	0.38
37.0	100	63	0.78
53.0	115	63	1.11
65.0	125	75	1.37
103.0	150	89	2.16

Note: Refers to nominal half round eaves gutters laid level with outlets at one end sharp edged. Round edged outlets allow smaller downpipe sizes.

Scenarios	Risk factor
Eaves gutters, where water overflowing would have little consequence	1.0
Eaves gutters, where overflowing water would cause particular inconvenience, such as at entrances to public buildings	1.5
Non-eaves gutters and where abnormal heavy rain, or a blockage in the system, could cause water to spill over into the building	2.0
Non-eaves gutters in buildings where an exceptional degree of protection is required, such as: • hospital operating theatres • critical communication facilities • storage intended for substances that evolve toxic or flammable fumes when wet • buildings housing outstanding works of art	3.0

Table 4.2 Risk factors for different design scenarios.

have been a number of climate change incidents that may need to be taken into account. The design rate categories are:

1. **For use with eaves gutters and flat roofs.**

   Return period of T = one year

   Rainfall intensities for a two-minute event shown in NB.1 (r for a two-minute storm and T = 1).

2. **For use with valley and parapet gutters, or where a building, or its contents, should have an additional measure of protection.**

   Return period of T = 1.5x L_y (probability of P_r = 0.5)

   Rainfall intensities for a two-minute event, during a storm of longer duration, shown in NB.2 to NB.5 (r for a two-minute storm and T = 5, T = 50, T = 500 years and Max probable r for a two-minute storm).

3. **For use where an even higher level of security than that provided by Category 2 is desired.**

   Return period of T = 4.5x L_y (probability of P_r = 0.2)

   Rainfall intensities for a two-minute event, during a storm of longer duration, shown in NB.2 to NB.5 (r for a two-minute storm and T = 5, T = 50, T = 500 years and Max probable r for a two-minute storm).

4. **For use where the highest level of security is required.**

   probability of P_r ~0.0

   Rainfall intensities shown in NB.5 (Max probable r for a two-minute storm).

The data from the rainfall intensity maps and the design rate categories are used in this probability equation:

$$P_r = 1 - [1 - (^1/_T)]^{Ly}$$

Where:

P_r = Probability

The probability of exceeding the chosen rainfall rate during the lifetime of the building.

0.0 = assured safety,

1.0 = certainty that the rate will be exceeded.

T = Return period (years)

The chance (i/T) that an event will be exceeded in any given year.

(Only applies if T is > five years).

L_y = Lifetime (Years)

Lifetime of protection for the building or its contents.

The selection of the design rate category appears to be rather arbitrary, based on perceived risk and required level of security, but the impact on the calculations can be significant. Designing for the maximum compared to the minimum design flow rate can introduce a factor of over 25, so the final design flow rate does need to be realistic bearing in mind the costs, appearance and actual implications of any overflowing of rainwater.

In some data rainfall intensity may be expressed in different units, for example:

$$r = \frac{(Dmin\ MT)}{D \times 60}$$

Although the units of mm/mm appear to be very different to litre/ (seconds. square metre), they are equivalent.

Rainfall and Climate Change

As climate change is now a factor in rainwater system design, BS EN 752 now includes a table of climate change allowance (CCA) multiplication factors that can be used to take account of future changes in the design rainfall intensity:

2. Effective Roof Area

The effective roof area, if no wind effects are considered is calculated as follows:

$$A = Lr.\ Br$$

Where:

A = Effective roof area (m²)

Lr = Length of roof, to be drained (m)

Br = Breadth of the roof, the plan length from the gutter to the ridge (m)

There are two options to determine the effective roof area if wind effects are to be considered. These are

End of design life	CCA (Central projection)	CCA (Upper end projection)
Up to 2039	1.05	1.1
2040–2069	1.1	1.2
2070–2115	1.2	1.4
Note: These recommendations are updated periodically in the light of new research. Users are advised to check for the most up to date guidance.		

Table 4.3. Recommended value of climate change allowance (CCA). This table is regularly updated. The latest figures can be found at: www.gov.uk/guidance/flood-risk-assesments-climate-change-allowances.

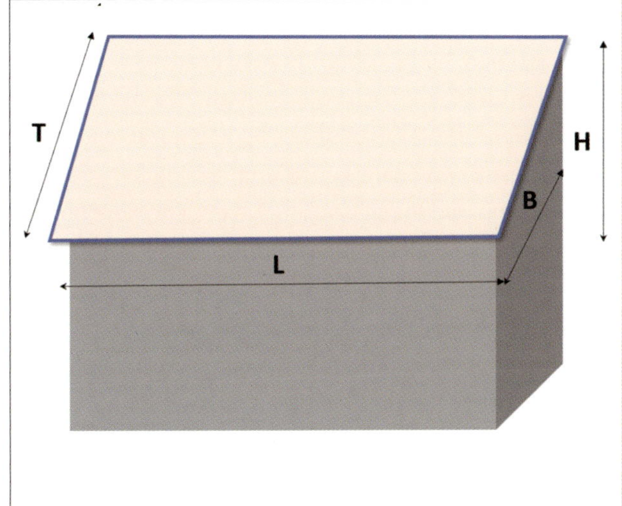

Type of allowance for wind effect	Effective roof area equation
Wind driven rain at 26° to the vertical	$A = Lr. \left(Br + \dfrac{Hr}{2} \right)$
Rain perpendicular to the roof	$A = Lr. Tr$
Where: A= Effective impermeable roof area (m²) Lr= Length of roof, to be drained (m) Br= Breadth of the roof, the plan length from the gutter to the ridge (m) Hr= Height of roof from gutter to ridge (m) Tr= Transverse distance from gutter to ridge, along the roof (m)	

Table 4.4 The two approaches for allowing for wind effect in roof.

summarized in Table 4.4. If wind effects are to be included, and where rain may be driven against a wall, the run onto the roof or into a gutter, then 50 per cent of the wall area should be added to the effective roof area.

3. Rate of Flow

The rate of water flow from the roof is calculated as follows:

$$Q = r.\ A.\ C$$

Where:

Q = Rate of flow of water from a roof (l/s)

r = rainfall intensity (l/(s.m²))

A = Effective roof area (m²)

C = Run-off coefficient (normally 1)

The design rainfall frequency is the *rainfall intensity that causes the pipe to be just full without surcharge.* Note that different design criteria may be set for combined and separate systems.

EN 752 provides a helpful table (Table 4.5) that compares different situations and suggested rainfall criteria to be used in the design. In addition, the National Annex includes a similar table for use within the UK but related to the established Risk Categories from EN12065-3.

Fig. 4.11a Typical flat roof with guttering-garage roofs.

Fig. 4.11b Doorway roof with felt lined gutter to a plastic hopper head.

Drainage in Flat Roofs

Flat roofs refers to roofs with pitches of less than 10 degrees. Thus, truly flat roofs do not really exist. Flat roofs can drain well during high rainfall but may perform badly during low-rainfall intensities, during which pooling of rainwater can occur and lead to local damage to the surface especially in freezing or high-sunlight conditions. Flat roofs will also be subject to sustained snow loading and may receive snow that falls from nearby sloping roofs, therefore, it is important to decide if this is the correct roof type for the building. If it is, such roofs should be detailed correctly and the appropriate slopes specified to avoid consequential damage to the roof. It is not unusual for rework to be required for flat roofs or that they are replaced by sloped roofs, as they can be less durable and require more maintenance compared to sloped roofs. Compared to pitched roofs, there are more options for draining water from a flat roof. These can include:

- Guttering
- A hopper through a parapet wall, or
- Rainwater outlet at any point within or at the edges of the flat roof.

Examples of guttering to low-pitch felt roofs are shown in Fig. 4.11.

In flat roofs, it is common for the lower edge to be

Location	Design Rainfall frequency [a]	
	Return period (years)	Probability of exceeding in any 1 year
Rural areas	1	100%
Residential areas	2	50%
City Centres/industrial/ commercial areas	5	20%
Underground railway/underpasses	10	10%
a - For the selected design rainfall event the pipe shall be no more than just full and shall be without surcharge		

Table 4.5 Example of design rainfall frequencies for pipes to run just full without surcharge.
(Source: BS EN 752 2017 Table 2 p.28)

Risk Category	Situation	Design rainfall frequency	
		Return period (1 in 'n' years)	Probability of exceedence in any one year
Category 1	Normal situations where ponding can be tolerated during heavy rainfall and for a few minutes afterwards	1	1
Category 2	Ponding cannot be tolerated	5	0,2
Category 3 (see note a)	Where a building or its contents require additional protection	1,5 x design life of building (see note b)	1/(1,5 x design life of the building (see note b)
Category 4 (see note a)	Where a building or its contents require a higher degree of security than category 3	4,5 x design life of building (see note b)	1/(4,5 x design life of the building) - (see note b)
Notes:			
a) Categories 3 and 4 should only be used in exceptional circumstances.			
b) Design rainfall intensities for other frequencies and durations can be calculated using the method described in BS EN 16933-2.			

Table 4.6 Recommended design rainfall frequencies for the design of areas around buildings. (Source: BS EN 752 2017 Table NA1 p.95)

Figs 4.11c-d Drainage chute and hopper for flat roofs.

Fig. 4.11e Flat roof drainage using parapet outlets with various hoppers and downpipes.

Fig. 4.11h Cascaded balcony and flat roof drainage utilizing bends, hoppers and downpipes.

Figs 4.11f-g Parapet rainwater outlet with emergency overflow clearly visible.

formed into a gutter profile, especially if a wall edges a balcony or accessible roof area. In these situations, the roofing, or flooring, material is used to form the gutter and an upstand produced that is attached to the wall by flashing to maintain water tightness.

The installation of a green or blue roof will reduce the run-off rate and have additional benefits, such as biodiversity, reduction in the heat island effect (in cities) and possibly the creation of an amenity space.

A balcony can be considered as a flat roof, but the amount of rainfall to which it will be subject depends on the design of the walls around it.

Modern Flat Roofs

Flat roofs are traditionally finished with roofing felt and topped with aggregate. The development of SuDS and urban greening projects has, however, led to the increased variations in flat roof design. Although such roofs may provide some additional insulation to a building, they are normally constructed with, or in addition to, insulation and water proofing layers. These new types of roof are named according to their prevalent colours:

Figs 4.12a–c Examples of variations of balcony drainage.

Fig. 4.13 Demonstrator of a brown roof.

Figs 4.14a–b A dwelling and garden shed using gently sloping green roofs.

Fig. 4.15a A blue roof structure using crates fitted with wicks to keep the green roof watered throughout the year.

Fig. 4.15b A typical green roof structure using granular soil to support the vegetation above the plastic water pathways on top of the base material. The correct type of soil is vital for the long-term survival and growth of the vegetation, hence the specification should make reference to BS 3882:2015 Specification for topsoil.

Fig. 4.15c Green roofs do not have to be on buildings; here is a bin store example.

Brown roofs are built without any vegetation, but may feature different types of aggregate to produce decorative features. They can also be made to resemble brown field sites and may use low-fertility substrates of rubble, gravel and rotten timber, which may lead to limited colonization by local vegetation.

Green roofs are vegetated roofs that can be either intensive or extensive. Extensive roofs simply have a covering of sedum, or grass or some other plant, which will develop a self-sustaining community of plants that require minimal maintenance.

Intensive roofs can provide amenity spaces such as patios and walkways, as well as large plants or trees. Hence, they need deeper soil and may require irrigation systems.

Blue roofs may or may not include vegetation, but are roof-mounted attenuation tanks for stormwater. A blue roof may be integrated into either a brown or green roof.

Although these types of roof are normally flat, especially extensive green roofs, they can be formed on sloping roofs if sufficient resistance to slippage is built in. All these roofs can be combined with solar panels and similar devices to generate power, but the surrounding plants should be compatible with any such equipment.

All of the above roofs will add to the loading of the building, hence the foundations, supporting walls and the roof supports must all be strong enough to withstand the additional construction and in-use loads.

Non-Conventional Systems

Traditional rainwater systems rely on the force of gravity to operate. As such, they work well under light rainfall, but during heavy rainfall the amount of water to transport can exceed their capacity and result in overflowing gutters. In most situations this is not a problem as the rain will be soaked up by the surrounding area anyway. However, if a valley gutter overflows in a factory, warehouse or other large building, the result could be a loss of stock or other internal damage. One way to increase the capacity of a rainwater system is to use syphonic action. Whereas a traditional system will tend to run at about one-third to two-thirds full, a syphonic system will run full and the resulting negative pressures can effectively suck a high volume of water along the collection pipes.

Although syphonic systems were originally developed in Scandinavia in the 1980s, they were not used in the UK until the end of the 1990s. Such systems are more beneficial for large buildings with flat roofs as they reduce the number of downpipes needed (often to just one) and do not require much roof space. Prob-

Fig. 4.16. Syphonic drainage pipe from a roof outlet, note the robust support bars to absorb the shock waves during the transition stages.

ably the most famous building in England with a syphonic rainwater system is London's Stansted Airport, but most retail 'sheds' built since then have fitted them. The disadvantage of syphonic systems is that while the pipework may be relatively inexpensive, the support structure for the pipes can be very expensive. The reason is due to the way syphonic systems work. When rainfall collects on a roof it will gradually run to the outlets and enter the rainwater system. The initial light rain will not be sufficient for the syphonic system to

Fig. 4.17 Syphonic outlet pipe configuration with fabricated trap.

run full, so it will act as a conventional system. During heavy rainfall, there will be sufficient rain entering the system for it to become 'primed' and syphonic action can take place. The syphonic action may be enough to empty the pipes of rainwater, and then syphonic action will cease until the pipes once again become filled. This process will continue throughout the rainstorm, resulting in frequent transitions to and from syphonic action. The transition phases result in large pressure fluctuations within the pipework and the generation of shockwaves. To prevent the pipework from blowing itself apart, robust supports are needed to absorb the energy and hold the pipework still.

The initial guidance and codes of practice that were developed in the 1990s have now been superseded by newer ones that have evolved due to knowledge and experience from many systems around the world. The latest standard is from the USA, ASPE/ANSI 45-2013: Siphonic Roof Drainage. In the UK guidance can now be obtained from the Syphonic Roof Drainage Association's website, www.siphonic-roof-drainage.co.uk. Guidance can also be found in BS 8490:2007 Guide to siphonic roof drainage systems.

At the end of a syphonic drainage system, the syphonic action has to be broken before draining by gravity into the local surface water sewer. If a combined sewer is being used, a trap should be specified to prevent odour transfer from the sewer.

Vacuum systems have also been used for rainwater drainage. Originally, the Centre Court at Wimbledon had a roof over the spectators that sloped toward the grass court. When it rained, the rain poured in front of the spectators and onto the grass. This soaked the audience and directed excess water onto the court. To collect the rain from the roof, conventional downpipes could be used, but they would have created 'restricted views' for some of the spectators. The solution, at that time, was to install a vacuum system. The rain from the roof was collected in discrete hoppers at intervals along the edge of the roof. The hoppers contained interface valves and once they were full the contents would be discharged upwards to the pipework under the top of the roof, from where the water would travel along to the collection tank near the main building. From there, it was disposed to the existing stormwater drain.

Chapter Summary

Rainwater drainage design depends on a number of factors such as location, situation, and shape of building and anticipated rainfall. In Europe, the design methodology is found in EN 12056-3:2000. Local building regulations may include aspects of EN 12056-3, but may only provide a speciation that is applicable to only very low levels of security. Examples of rainwater system design are provided.

There are wide ranges in the shape and material of guttering and downpipes. Purchase and installation costs can influence choice, but as most gutters are exposed, aesthetics are also an important consideration. In addition to traditional rainwater systems, syphonic and vacuum systems can be used in appropriate circumstances.

Chapter 5

INTERNAL DRAINAGE DESIGN

Drainage systems internal to a building play an important role in maintaining an effective and efficient drainage train. As discussed in the previous section, the internal drainage system collects wastewater at the point of use to transport this onwards to an external drainage system. Alternatively, the internal drainage system can facilitate the recycling and reuse of the rain and greywater within the building to improve water use efficiency and minimize unnecessary wastewater discharge into the sewer system, without compromise to the effective function of the sewer network.

Key performance functions of internal drainage systems include:

- To transport waste and foul water safely from appliances used for food preparation, cooking and washing in a building to a public or private sewer, septic tank or cesspool. With the exception of water or wastewater diverted for reuse
- The removal of water that has been contaminated with waste products; and
- The provision of a physical barrier between the potentially harmful odours and fumes present in drainpipes and sewers and the habitable space (Jack and Swaffield, 2009).[1]

The types of wastewater generated from water use in a building include: domestic water, trade effluent, grey-water and blackwater. The type of wastewater informs the drainage system used for its handling, management and disposal.

The most relevant standards are BS EN 12056:2000 for Gravity drainage systems inside buildings – Parts 1–5, and BS EN 12109:1999 for Vacuum drainage systems inside buildings.

BS EN 12056:2000 provides the following important definitions to inform the choice of, and design of, any internal drainage system:

- **Wastewater:** water contaminated by use and all water discharging into the drainage system, e.g. domestic and trade effluent, condensate water and also rainwater when discharged into a wastewater drainage system
- **Domestic wastewater:** water that is contaminated by use and normally from WC, showers, baths, bidets, wash basins, sinks and floor gullies.
- **Blackwater:** wastewater containing faecal matter or urine
- **Greywater:** water not containing faecal matter or urine
- **Trade effluent:** water after industrial use and processes; contaminated/polluted water including water used for cooling.

Greywater and blackwater drainage inside the building can be pressure, vacuum or gravity conveyed.

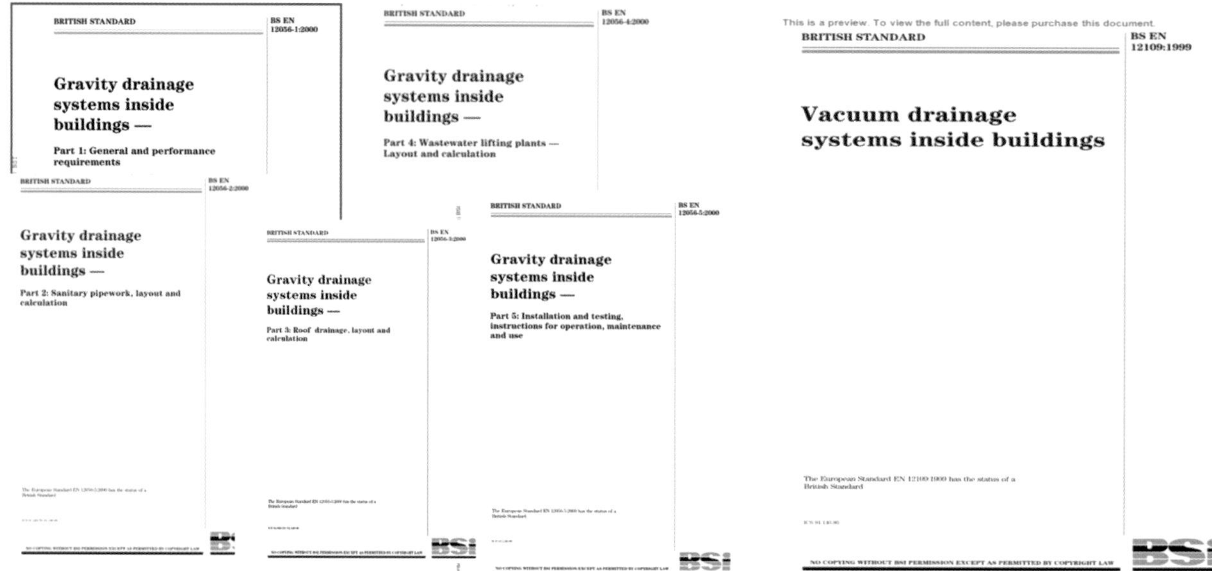

Fig. 5.1 Notable BS standards for internal drainage design.

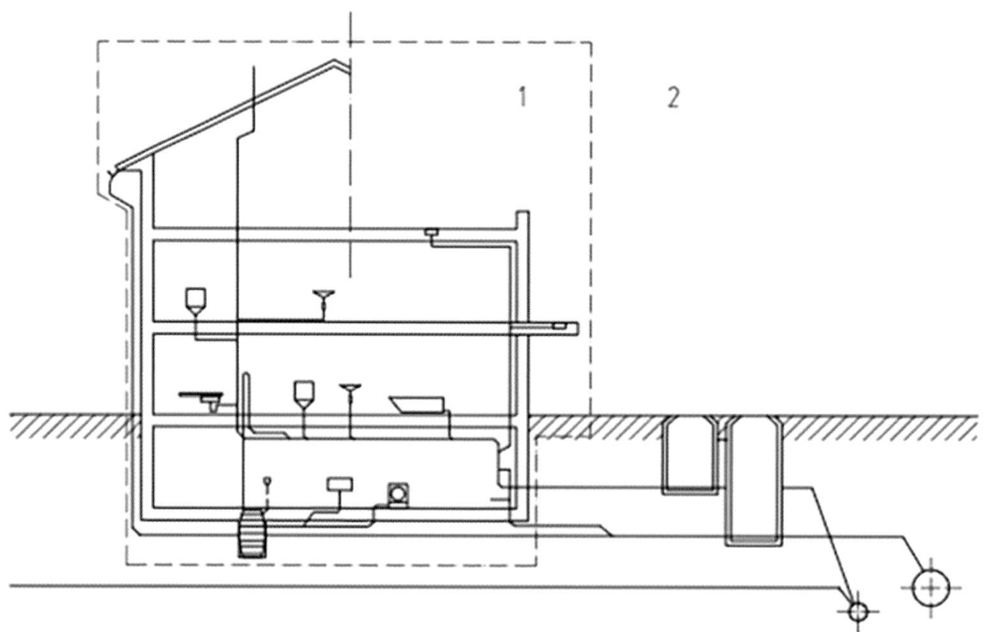

1 Gravity drainage systems inside buildings
2 Gravity drainage systems outside buildings

Fig. 5.2 Scope of application of internal and external drainage systems.
(Source: BS EN 12056: (2000) Gravity drainage systems inside buildings)

Like external drainage systems, internal drainage systems also have two main types:

- **Combined system:** This is where the same drainage assembly is used for wastewater i.e. greywater and blackwater foul drains. This system needs to be ventilated at the top of each main discharge stack.
- **Separate systems:** Here there is a separate drainage assembly for blackwater such as WCs and associated washbasins, compared to other forms of wastewater. Separate systems are more prone to problems with deposits from surfactant-using appliances as there is less dilution from WC flushes.

Design Considerations

The foul water drainage system of the building falls under mechanical systems considerations. In England, the design performance criteria for mechanical systems are covered under Parts 6–9 of the building regulations and more relevantly Parts G: Sanitation, Hot Water Safety and Water Efficiency and Part H: Drainage and Waste Disposal of the Schedule 1 requirements (and their approved documents thereof) of the said regulations. These highlight that drainage and wastewater systems cannot be designed effectively without a full understanding of water supply and use systems within the building.

Useful standards and guidance documents include:

- Building Regulations Approved Document H
- BS EN 12056 (2000) Gravity drainage systems inside buildings
- BS EN 12109 (1999) Vacuum drainage systems inside buildings
- BS EN 806 (2006) Specifications for installations inside buildings conveying water for human consumption (Some parts under review)
- BS EN 274 (2002) Waste fittings for sanitary appliances (Under review)
- BS 8533 (2017) Assessing and managing flood risk in development. Code of practice.
- BS 8525 (2010) Greywater systems. Code of practice
- BS 4660 (2000) Thermoplastics ancillary fittings of nominal sizes 110 and 160 for below-ground gravity drainage and sewerage
- BS EN 1057 (2006+A1:2010) Copper and copper alloys. Seamless, round copper tubes for water and gas in sanitary and heating applications
- CIBSE Guide G (current version): Public Health & Plumbing Engineering.

Broadly, any internal drainage system design should consider the following:

- **Spatial functions:** It is important to consider the optimal design and location of internal spaces that require water supply and drainage, e.g. toilets and bathrooms, kitchens, cloakrooms, utility spaces, etc. These spaces should be stacked in multi-storey buildings. Ill-considered and random location of these spaces could result in too many branches and unnecessarily staggered and untidy pipework, which is prone to blockage and other problems. In addition, spatial provisions should be made for plant and equipment, especially for non-gravity systems or where recycling and reuse systems are specified. Architects typically underestimate how much space is required for mechanical systems in buildings, especially as building systems become more complex and sophisticated.
- **Number of floors:** Drainage is simple in single-storey buildings as there are no stack effects to consider. Rooms or appliances may simply drain to gullies outside the building or directly to the drainage system. Once a second floor is introduced, the discharges of one floor will affect the discharges of the other. The simplest solution is to drain the upper floor by a different means to the ground floor. For multiple-floor buildings, consideration needs to be given to the anticipated use of the

building and its drainage system.

- **Water supply, sanitary appliances and plumbing fittings:** Drainage systems are underpinned by an understanding of the projected water use in buildings as well as the various water using appliances and fittings in a building. Lack of understanding and design resolution could result in an under- or over-specified pipework.
- **Layout:** Further, functional spaces and layout of the building should be such that it enables the pipework to be as simple as possible, and to minimize changes in direction and gradient.
- **Pipework capacity:** The capacity must be sufficient to carry the expected flow at any point of use in the building. This is determined according to the:
  - Volume of discharge
  - Flow rate
  - Gradient
- **Access for maintenance and repairs:** Systems should be designed to permit access for maintenance and repairs as well as adaption, extension and modification to meet future needs. Drains are prone to blockages due to poor design, installation or use. Therefore, it is important that they have appropriate, sufficient and suitably located access points in drain runs. This should include access at the head of each drain run, at a bend or change in gradient, change in pipe size and at a junction.
- **On-site alternative supply/renewable systems:** the opportunities and impact of recycling and reuse of wastewater in a building should be considered during early drainage design to avoid potential problems as a result of inadequate water flow, improper sizing or cross-contamination in the system. However, these problems are avoidable if considered early in the design process and should not be used as an excuse for not undertaking innovative water solutions in buildings.
- **Other risk factors:** Risk factors such as overflow, backflow or surcharging, for example due to flooding, should be considered in drainage design. This is currently not an explicit requirement in the building regulations. However, it is covered in other regulations such as the Flood Reinsurance Regulations Act 2015 etc. However, this does not negate this as a design responsibility, especially if the building is sited in a high-risk area such as a flood plain. The BS 8533 (2017) and the Approved Document H – Drainage and Waste Disposal 2015 provide guidance on flooding risks in external drainage. Similarly, internal drainage should be designed to avoid backflow in the network or internal flooding either due to excessive pressure, failures of connections, faulty appliances and – as much as is feasible – taps and appliances left running and forgotten.

In view of these, the design of internal supply and drainage pipework should consider the following:

- Points and amount of water demand and waste removal.
- Capacities for hot and cold water systems (storage, pumps, etc.). Include other sources, e.g. rainwater harvesting.
- Plant location and sizes, including water storage, waste storage.
- The main pipe and drain routes from floors to and from risers. Include consideration of other supply sources, e.g. rainwater harvesting.
- The main below-ground drainage routes and manhole locations so that above and below-ground drainage can be coordinated with the rest of the design: superstructure (walls/floors) and substructure (foundations). Loading from structure and other elements should be considered.
- Noise from the use of appliances and from the pipes during use.
- Material selection for durability, frost protection and corrosion resistance. Also, to avoid condensa-

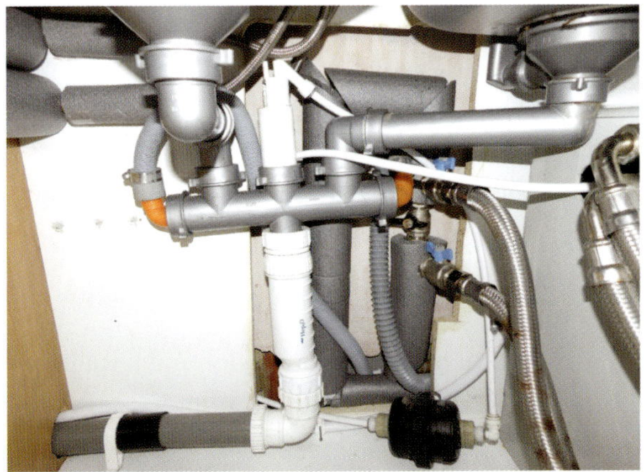

Fig. 5.3 The vertically mounted membrane valve trap uses a conventional manifold pipe to drain two sinks, a washing machine, dishwasher and the rinse water from an adjacent water softener. This maximizes the cupboard space and maintains access to the various water isolation valves. The overflow pipe from the water softener is accommodated through the flexible white tube that runs along the main waste pipe to the outside.

tion.

- Conflict with other services. Sanitary discharge pipes may sometimes need to be accommodated in spaces that will be shared with water supply pipes, electrical outlets and isolation valves, which all need to be accessible. The use of compact or membrane valve traps that require less space may be a viable option. However, this may result in a complex assembly as shown. So trade-offs may be needed (Fig. 5.3).

There are a number of unknowns that should also be considered. For instance:

- Usage factors for appliances and fittings
- Intervals between discharges into the system
- Materials and items that users will put into the drainage system.

The determinable and assumed factors are summarized in Table 5.1.

Known (determinable) factors)	Unknown factors
<ul><li>Type and number of appliance or fitting</li><li>Pipe size and fill</li><li>Pipe material: durability, etc.</li><li>Ventilated or unventilated</li><li>Topography of site</li><li>Location of sewers/tanks/pits /receiving waters etc.</li><li>Steady state performance</li><li>Transient performance</li></ul>	<ul><li>How much the appliances or fittings will be used</li><li>Intervals between discharges into the system</li><li>Materials or items to be put into drainage system by users:<ul><li>- **Deposition**</li><li>- **Blockage**</li><li>- **Damage**</li></ul></li></ul>

Table 5.1 Known and unknown design for internal drainage systems.

Design components

The main components of an internal waste and foul water drainage system include:

- **Discharge pipes:** These typically convey wastewater from the source, i.e. the appliance or fitting, to the drainage system.
- **Water appliances and fittings:** Appliances are important components of drainage design. They dispense, contain or use water for washing and cleaning. Examples include: wash basins, baths and showers, washing machines and dishwashers, toilets and urinals.
- **Discharge stacks:** This mostly consists of vertical and branch pipes that are used to transfer foul water from kitchen and sanitary appliances in

buildings. These pipes are typically grouped together as stacks that run vertically through the floors of the building to avoid unnecessary complexity and problems such as blockages in the system. This is why toilets and bathrooms are situated in the same vertical location in multi-storey buildings. Discharge stacks must be properly sited relative to the location of appliances and fittings as well as connections to the on-site and off-site drainage system. In addition, discharge stacks can be ventilated or unventilated depending on the type of internal drainage system.

- **Ventilating stack:** Typically connected to a discharge stack to limit pressure fluctuations.
- **Air admittance valves:** Mechanical valve that allows ventilation air to enter a system, but will not allow air to leave it. Can be fitted to a stack or branch or appliance.
- **Traps and siphonage:** Water traps are seals, which retain water in the pipe. They are fitted to foul drains to prevent the odour travelling back up the pipe.

Gravity Drainage System

The general recommendation is that all drainage systems installed above the flood level should operate by gravity and do not need to be connected to anti-flooding devices. Other sanitary installations, for instance in basements, can be pumped or use other wastewater lifting methods and should be drained via anti-flooding devices.

Pipe Types, Sizing and Gradients

Smooth rigid PVC-U pipes, with appropriate connections, are the most common materials used for internal drainage pipework.

Table 4 of the Building Regulations Approved Document H states that cast iron, copper, galvanized steel, and plastics such as Polypropylene (PP), ABS, Polyethylene (PE) and PVC-U, PVC-C and PVC+SAN can be used as appropriate for the type of waste discharge. For instance, not all materials will be appropriate for trade effluents. Also, non-metallic materials may be used to connect different metal types to avoid electrolytic corrosion, and all extraneous-conductive metal pipes should be adequately earthed.

Suitability of materials, connection and installation techniques should be checked according to the following:

- National regulations e.g. the building regulations in the UK
- Local, national and/or international technical standards and approval processes. E.g. the British, EU or ISO Standards e.g. BS EN 1329, 1565, 3868, etc
- Quality Assurance Schemes and Product Certification Schemes. The UK example is the British Board of Agrément (BBA) Certification
- *Conformité Européene* (CE) marking under EU Directives and Regulations or other Construction Products Regulations

It is important to design internal drainage to have enough capacity to carry the amount and flow of wastewater within the system. The flow, diameter and gradient of the pipe is calculated based on the number and capacity of appliances and fittings connected to the system.

NUMBER OF DWELLINGS	FLOW RATE (LITRES/SEC)
1	2.5
5	3.5
10	4.1
15	4.6
20	5.1
25	5.4

** Typical house flow based on: 1 × WC, 1 × bath, 1 or 2 washbasins, 1 × sink and 1 × washing machine.*

Table 5.2 Flow rates from dwellings.

Peak flow (litres/sec)	Pipe size (mm)	Minimum gradient (1 in...)	Maximum capacity (litres/sec)
<1	75	1:40	4.1
	100	1:40	9.2
>1	75	1:80	2.8
	100	1:80*	6.3
	150	1:150†	15.0

Notes:
* Minimum of 1 WC   † Minimum of 5 WCs

Table 5.3 Recommended minimum pipe gradients (Source: Table 6 Building Regulations Part H 2015)

Drainage Stacks and Branches

Discharge stacks are vertical drainage pipes used to group and align the drainage pipework of sanitary appliances in buildings, especially in buildings with multiple storeys.

The BS EN 12056-2: 2000 (Gravity drainage systems inside buildings) defines four main classification of internal drainage system types:

System I. **Single discharge stack system with partly filled branch discharge pipes** where a branch that is partly filled to a filling degree of 0.5 (50 per cent) is connected on one end to appliances and on the other to a single discharge stack.

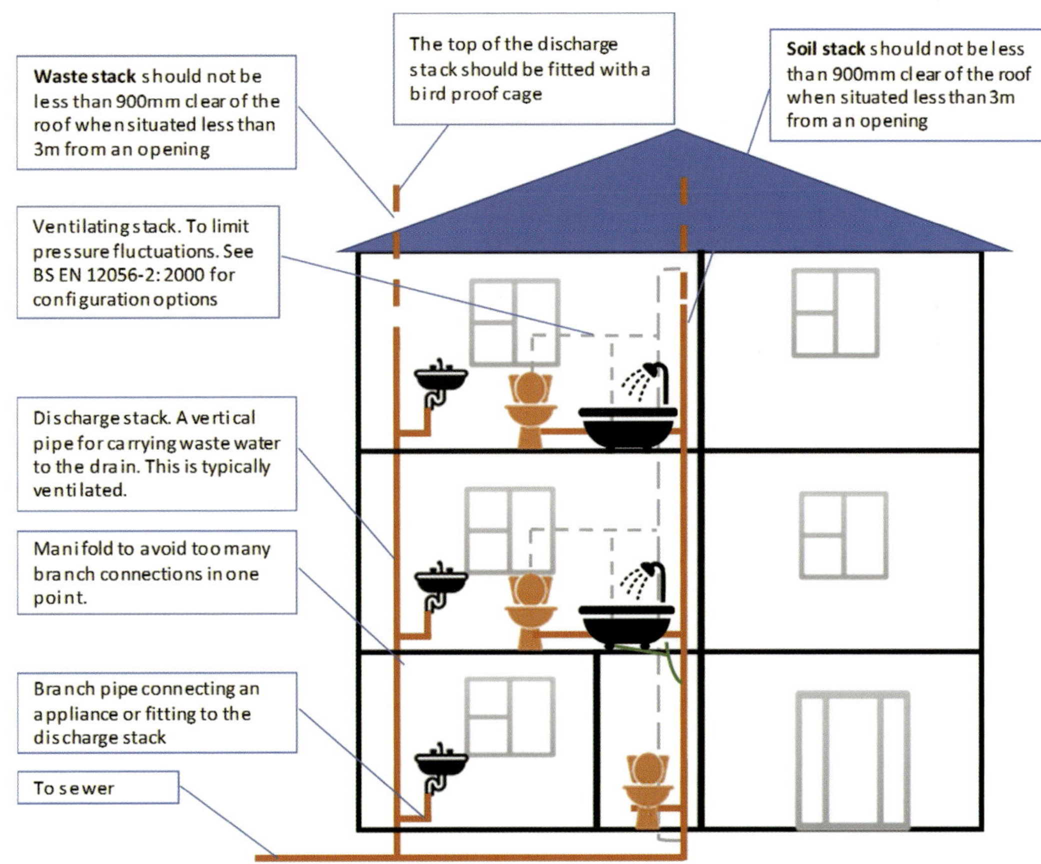

Fig. 5.4a Example configuration of a primary ventilated discharge stack system serving basins, with a secondary ventilated stack serving the baths and WCs.

System II. **Single discharge stack system with small-bore branch discharge pipes** where small-bore branch discharge pipes designed with a filling degree of 0.7 (70 per cent) are connected on one end to appliances and on the other to a single discharge stack.

System III. **Single discharge stack system with full-bore branch discharge pipes** where full-bore branch discharge pipes designed with a filling degree of 1.0 (100 per cent) are connected on one end to appliances. However, each branch discharge pipe is connected separately to its own single discharge stack.

System IV. **Separate discharge stack system** can include types I, II and III but the types of wastewater, e.g. blackwater from WCs and urinals, and greywater from other appliances, are separated for reuse. This system is also used where sanitary appliances are located a significant distance apart. Although, for cost reasons, this may not be necessary and could be avoided through good spatial design and layouts. In this instance, only the blackwater drain may need to be ventilated using a stack. However, there are many examples where the drains are ventilated in a manner similar to that of a combined system.

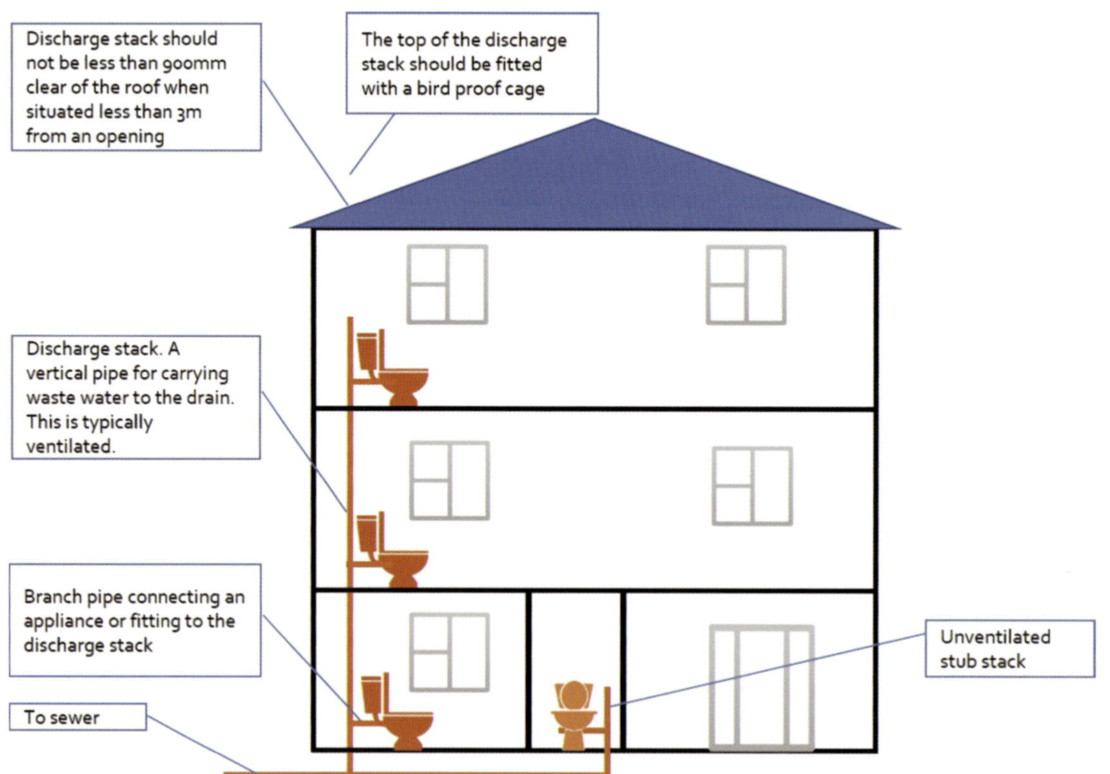

Fig. 5.4b Example of a primary ventilated stack with a stub stack.

Fig. 5.5d Example of a poor externally located sanitary discharge pipework system. Leaks are clearly visible.

Fig. 5.5a Photograph of single-storey ventilated stack on the external façade of the building (centre) and a rainwater downpipe with hopper head (to the right). Note how the ventilation pipe is offset around the window and terminates above the roofline.

Figs 5.5b–c Photograph of a clean multi-storey drainage stack on the external façade of the building.

Figs 5.5e-f Example of French drainage layout, within a shaft and with range of gradients for the branch pipes.

Figs 5.6a-b With an internal drainage system, often the only visible part is the air inlet to the ventilation or soil and vent stack. However, beneath the tiles the stack runs through the loft space with an offset.

Design Stack Type and Configuration

Calculations based on discharge units and flow rates per appliance will be conducted by the M&E engineer in accordance to guidelines. Four main configurations types are used separately or combined, each with some variation on how they deliver pressure control:

1. **Primary ventilated systems:** pressure control is achieved by airflow in the discharge vent, discharge stacks or air admittance valves.

2. **Secondary ventilated system.** Pressure control is achieved by the addition of a separate ventilation

Fig. 5.6c Example of external discharge pipe and inspection chambers, enabling access to WC branches and access chambers without entering the building.

Fig. 5.6d External drainage pipework in ventilation shaft of the London underground.

stack and/or secondary branch ventilation pipes connected with the stack vents.

3. **Unventilated discharge branch configuration.** Pressure control is achieved by airflow in the discharge branch.

4. **Ventilated discharge branch.** Pressure control is achieved by ventilating the discharge branches or using air admittance valves. Examples as previously shown.

The key thing to note from an architectural point of view is that the principles of these systems result in stacks that can be configured in multiple ways to manage pressure within the system but also suit design and functional needs in buildings. Different configuration options can be proposed to suit domestic versus non-domestic buildings or single or multi-storey buildings. In multi-storey buildings, e.g. apartment blocks for ex-

ample, it is good practice to locate sanitary spaces such as toilets and bathrooms in the same locations on each floor so that the drainage from these fittings can run along the same vertical stack (Fig. 5.7).

Internal pipework is not generally visible from outside the building. Access to internal pipework may be located externally to avoid disruption and possible contamination of the inside of the building. Where internal drainage runs cannot be avoided, double seal access chambers should be used. These normally have bolted down profiled pipe covers within a conventional framed chamber that has a cover sealed using a grease seal or a no-less-effective airtight closure. A ventilation stack is also added to any branch that is more than 6m long serving one appliance or 12m long serving a group of appliances.

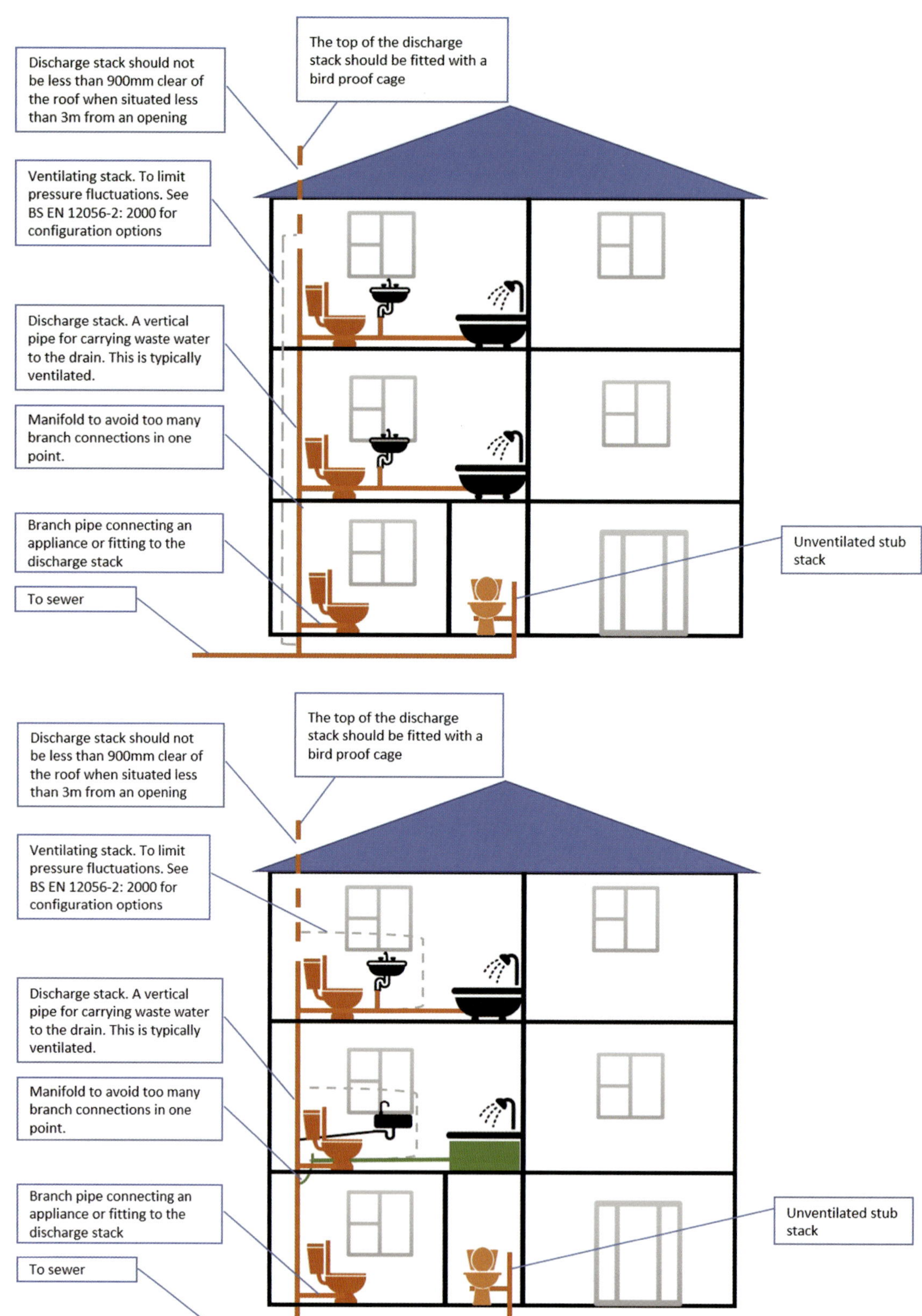

Figs 5.7a–b Examples of secondary ventilated stack configurations.

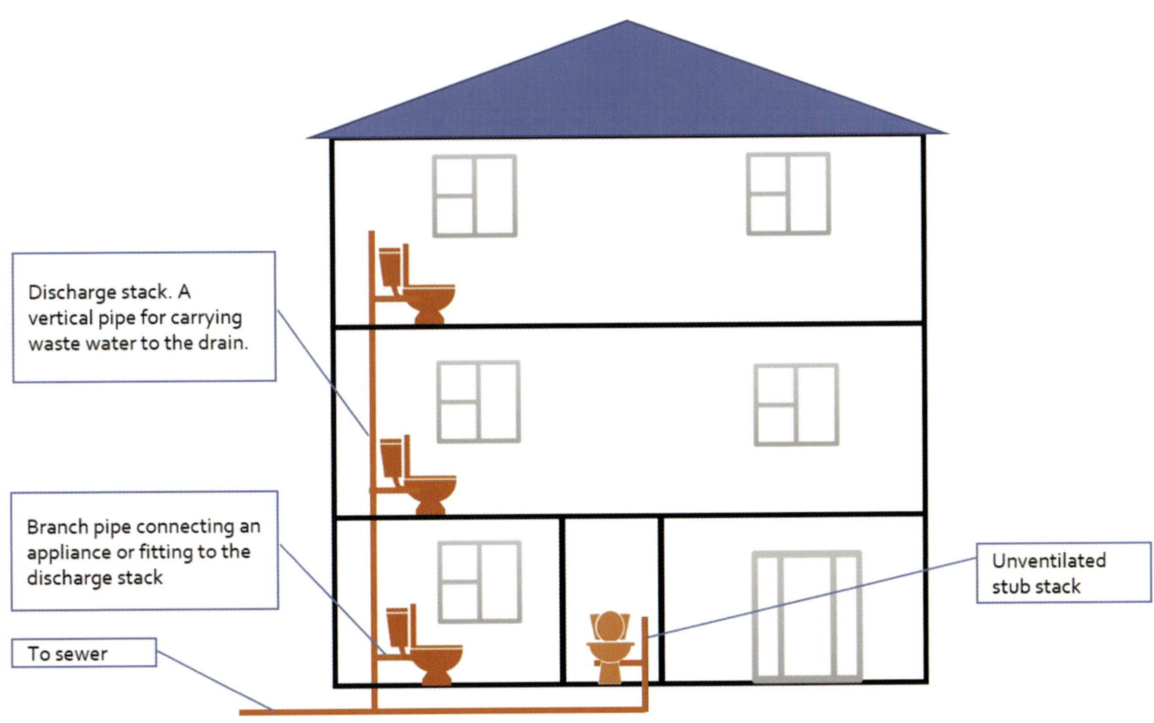

Fig. 5.7c Another example of an unventilated stack configuration.

Why Use Ventilated Stacks?

With sewerage design, typical flows at continuous and 'self-cleansing' velocities are key to minimizing deposition problems. However, very few drainage systems experience continuous flows. The flows tend to be intermittent, variable and difficult to predict. Hence, much design is based on the mathematics of probability. This results in design principles or rules that will contain 'safety factors', usually of a varying percentage, that take into account the likelihood of extreme situations and flows.

However, there are instances where sections of drain with different specifications are installed close to each other or are needed to perform beyond their normal design limits. For example, if a stub stack is connected to a very deep drain of, say, 3m below the branch inlet, there should not be any problem unless an extreme load is flushed (e.g. a 'disposable nappy') and a drain surcharge occurs at the same time.

A typical pressure profile for a discharge from a branch into a stack is shown. This shows that both positive and negative pressures will be generated; negative towards the top of the stack and close to the point of discharge, but positive close to the bottom of the stack where the air flow is cut off. Notice that the magnitude of the negative pressures is far greater than the positive pressure for this situation.

However, the pressures generated within a completely unventilated stack, such as a stub stack, would produce a different profile. In a normally ventilated stack pipe, air can enter the top from the atmosphere, so the water falling down the stack, in an annulus, can reach its terminal velocity due to the resistance of the falling water on the rising air and the surface of the stack. However, when the stack is closed to the atmosphere no air can enter the stack above the discharging water, so the discharge tends to form a plug of water, like a piston. As the water falls, the drop in pressure above the water creates a suction that slows down the falling water. At the same time, the plug of falling water is very effective at compressing the air below it and if there is an offset or drain surcharge the positive pressures generated can be significant (this is one reason that stub stacks are limited in height in building regulations). This is the reason why ventilated stacks are typically included in any internal pipe network to relieve excessive pressure within the pipes.

Air Admittance Valves

A vent stack may be terminated within the roof space and the air flow controlled by a valve. An air admittance valve is used instead of an open stack, that needs to penetrate roofs and is subject to wind effects. The valve is normally held closed by gravity or a spring, however, when a suction is applied to the inside of the valve it will open to admit air. Once the pressure inside the valve has increased to atmospheric, again, the valve will close. If the pressure inside the valve rises above atmospheric, the valve will stay firmly closed, preventing any release of odours.

Although many such air admittance valves (AAVs) are installed on main ventilation stacks, they can also be fitted to branch stacks, branch discharge pipes and even appliances. Generally AAVs are fitted well above any flood level in a system, but if they are fitted to traps or pipes beneath basins or sinks a more robust type of valve will be required.

The European Standard (EN 12380: 2002) includes performance requirements for two types of AAV,

above (Class B) and below appliance flood level (Class A), as well as three different temperature environments. The temperature environments relate to different locations in which the valves may be located:

I. **-20 to +60°C**; for installations in uninsulated roof spaces that are subject to sub-zero winter temperatures and very high summer temperatures.

II. **0 to +60°C**; for installations in areas that are not subject to sub-zero winter temperatures, but are likely to experience very high summer temperatures.

III.**0 to +20°C**; for installations in well insulated roof spaces that are not subject to sub-zero winter temperatures or high summer temperatures.

Standards for AAVs are also available in the USA and other countries. These include:

- **ASSE 1050** Performance Requirements for Stack Air Admittance Valves for Sanitary Drainage Systems.
- **ASSE 1051** Performance Requirements for Individual and Branch Type Air Admittance Valves for Sanitary Drainage Systems.
- **AS/NZS 4936:2002** Air admittance valves (AAVs) for use in sanitary plumbing and drainage systems.

Although AAVs are now common on above ground drainage systems, they only deal with the negative pressures generated within the drainage stack. To complement the AAVs or simply to relieve positive pressures without using a vent pipe, a positive air pressure attenuation (PAPA) device has been produced (Fig. 5.11). This item absorbs the excess positive pressure using a resilient membrane. The units can be joined together to provide greater attenuation if needed. They can also be located at intervals along the stack so that systems serving buildings with over forty floors can be adequately ventilated using a combination of PAPAs and AAVs.

Traps and Siphonage

A trap is used to prevent foul air and odour from

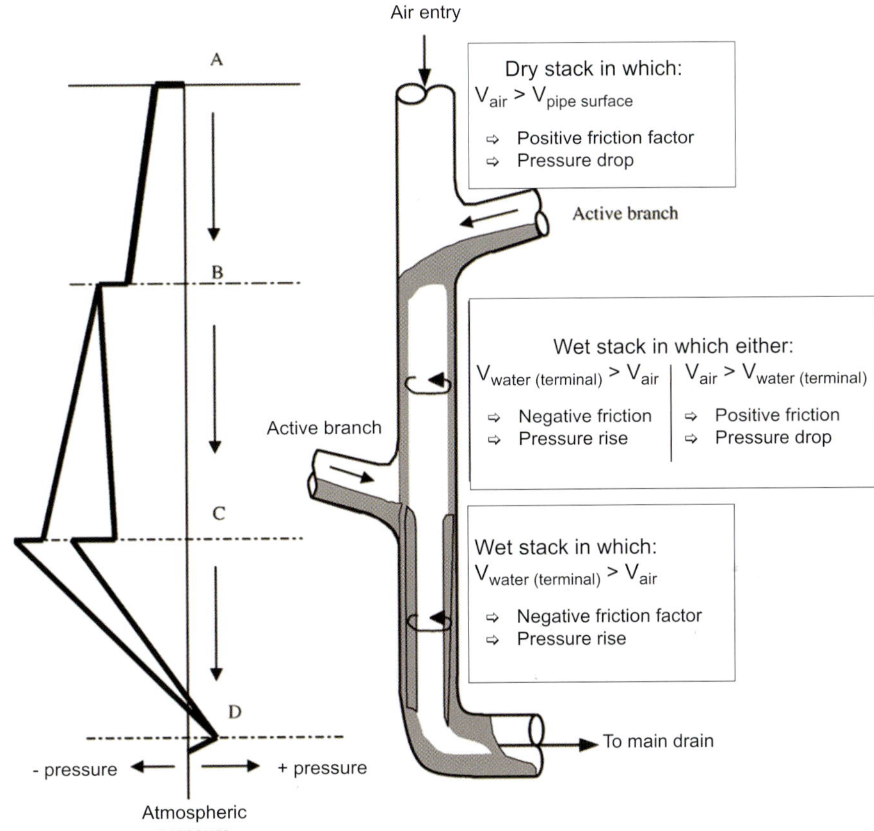

Fig. 5.8 Air pressure fluctuation down a typical multi-storey stack with a discharge from a higher branch.

Fig. 5.9a AAV installed, on a branch stack, above WC flood level within a toilet cubicle.

Fig. 5.9b Stack AAV installed above the flood level of the basins and urinal.

Fig. 5.9c Branch AAV installed above the flood level of the urinal. Note that the positioning of these stacks is not ideal and can be considered aesthetically unsightly.

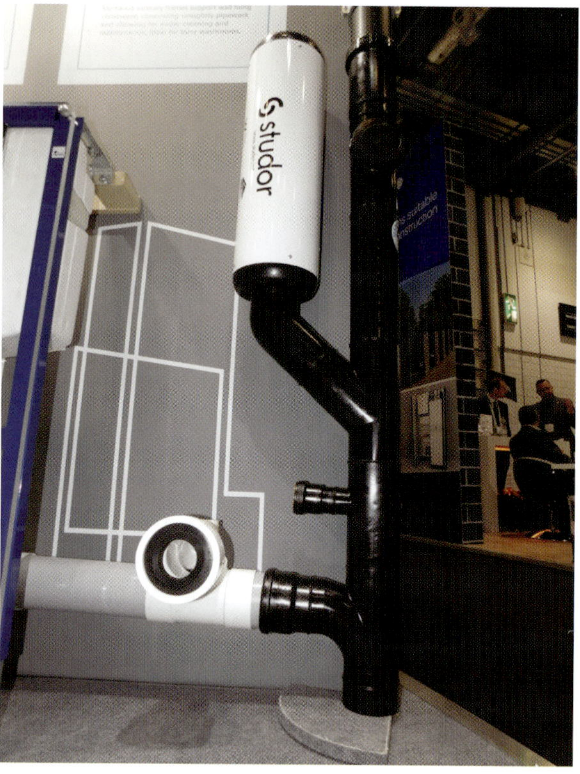

Fig. 5.10 A PAPA on demonstration attached to drainage stack.

waste/foul water pipes penetrating into spaces within the building. In instances where there is persistent odour from the drainage system it is often necessary to check that both the traps and stacks are designed and functioning properly. Detailed guidelines for traps can be found in BS EN 274. However, most building regulations require that a trap is installed to all sanitary appliances, especially gullies and WCs, with the rest of the system designed to prevent siphonage or any other displacement of the trap seal.

Siphonage occurs when the water in the trap is drawn out by the water flowing through the pipework.

- Self-siphonage can occur in pipes and stacks if they run 'full bore', e.g. a WC being flushed or bath being emptied; as the momentum of the flow draws the water seal from the trap.
- Induced siphonage is the result of a drop in air pressure in the trap outlet caused by a discharge from one or more nearby appliances. The reduced pressure draws out water from the seal of the affected fitting.

Therefore, ventilation/anti-siphon pipes and stacks are added to limit the fluctuations of air pressure so that trap seals are maintained. The limits for UK drainage systems are normally ±38mm water gauge ($325N/m^2$, 325Pa,) These have been derived from imperial units,

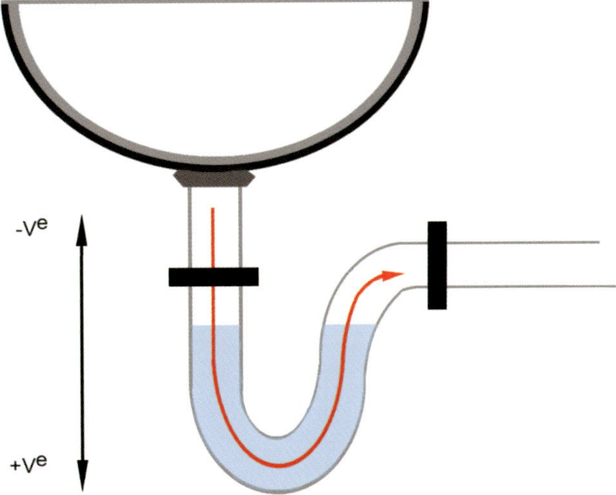

Fig. 5.11 Pressure fluctuations in pipes.

where a 1.5in tolerance on a 3in depth of seal trap was considered to be a reasonable level of performance. A 1.5in water gauge equates to 38.1mm, so 38mm is roughly equivalent, bearing in mind the issues of reading such a value with a meniscus in a tube.

The design and layout of an above-ground drainage system will include a number of features that can influence the air pressure changes within the system. These include:

- Oversized stacks, fitting branch appliances too close to the base of the stack. The height of the lowest discharge branch should be a minimum of 450mm for up to three storeys, and a minimum of 750mm for up to five storeys. A separate stack is also required for lower-level appliances in buildings with more than five storeys
- Opposed branch connections without the required minimum offset or a manifold
- Offsets in the wet part of a drainage stack (there are no problems with offsets in the dry parts of stacks).

The main issue with the above features is the intermittent water curtain, or curtains, that can form during a discharge. While water falls down a vertical stack it will adhere to the sides and fall as an annulus with air flowing upwards in its core. However, at a bend or junction the falling water will create a curtain that reduces air flow. With an offset, where there are two bends, the double water curtains can have a much

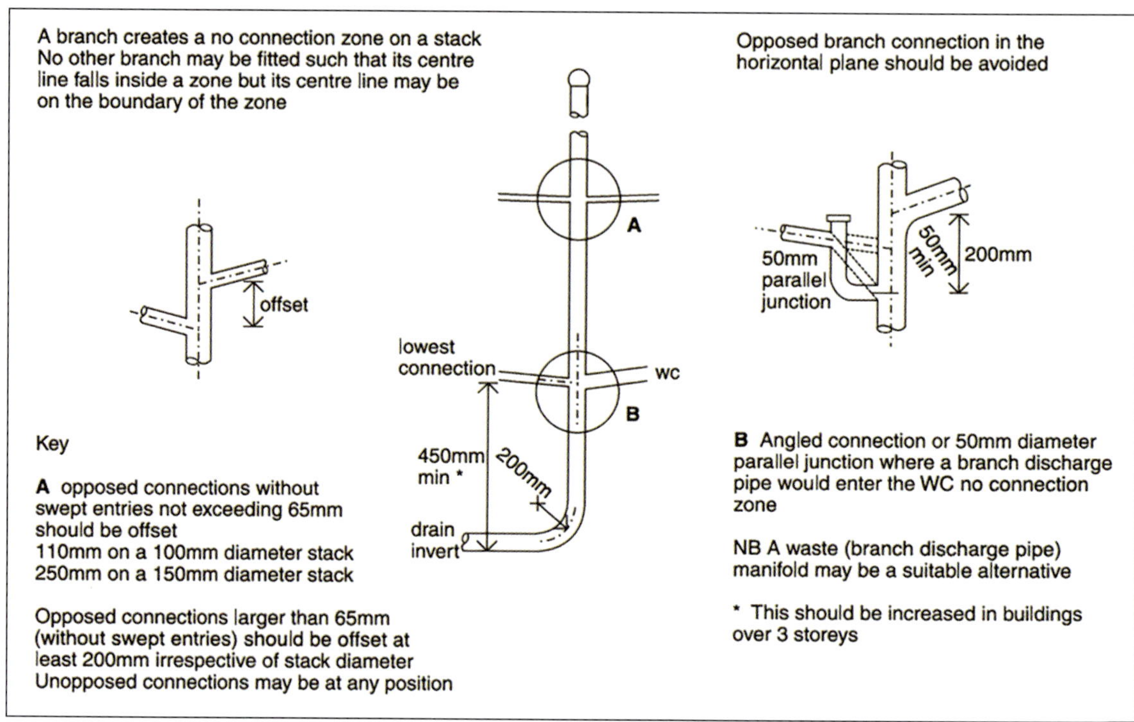

Fig. 5.12 Crossflow prevention in branch connection to stacks. (Source: Diagram 2. Building regulations (2010) Approved Document H)

greater impact as they make a more effective barrier for the airflow. So, if the discharge is long enough – in time and distance – to create water curtains in both bends at the same time, bypass ventilation may be needed if other appliances will be affected.

In addition, any blockage in a trap, or pipework, due to items flushed down the drain or a build-up of deposits, can limit air flow. High-foam detergents can be a problem in stacks that only serve laundries or bathrooms.

There are now a few innovative traps in the market that can be specified to alleviate some of these problems, including self-resealing, ventilated, valve-controlled anti-siphonage traps.

Common trap terminology is as shown in Fig. 5.13.

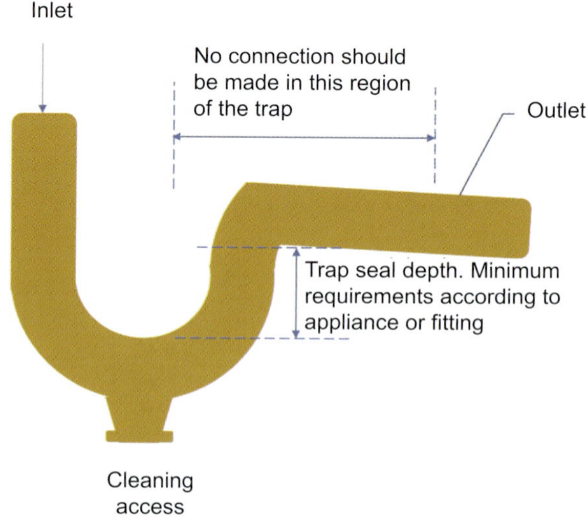

Fig. 5.13 Trap terminology.

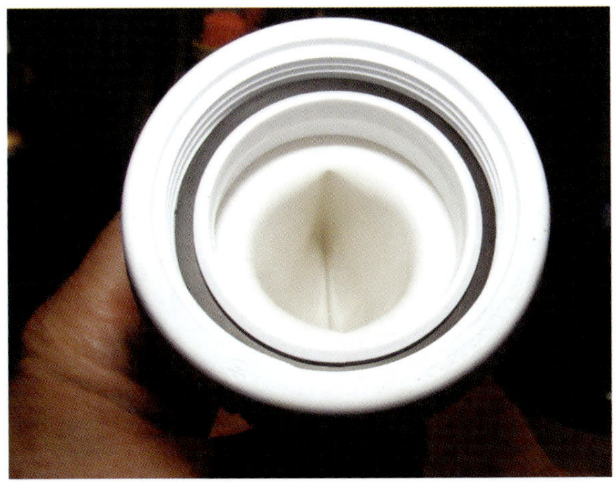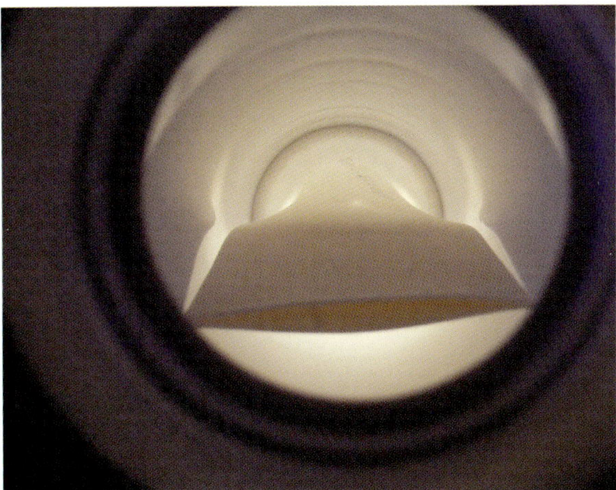

Figs 5.14a-b The inlet and outlet of a typical membrane trap. When water flows into the trap the tubular membrane opens and passes the water. As soon as the fallow ceases the membrane closes, producing an airtight seal to prevent the passage of odours.

Traps are commonly named according to their shapes, e.g. the 'S' trap, bottle or 'P' trap (Fig. 5.13). As shown, the outlet from the S trap is vertical, while the P and bottle trap outlets should be installed with a slight inclined dip to the horizontal.

Other important criteria are the trap size and depth of water seal.

The depth of water seal varies with appliance and location. In the UK traditionally a 75mm depth of seal was used for most appliances with separate traps. However, many European systems with 50mm seals were normal. Today some manufacturers are catering for global markets and producing traps with 80mm depths of seal.

For appliances such as WCs, a 50mm depth of seal is normal, but low-volume flush WCs sometimes use only 25mm seals protected by attached air admittance valves. For ground-floor appliances [where pressure fluctuations due to other appliances that are connected to separate stacks will not impact them], or appliances discharging into external hopper heads, a 25mm depth of seal is often used. This is typical in any installation discharging to a trapped gully as the gully trap will

Fig. 5.15 Example set-up of compact systems used in basements.

Fig. 5.16 P Traps from a double-bowl sink.

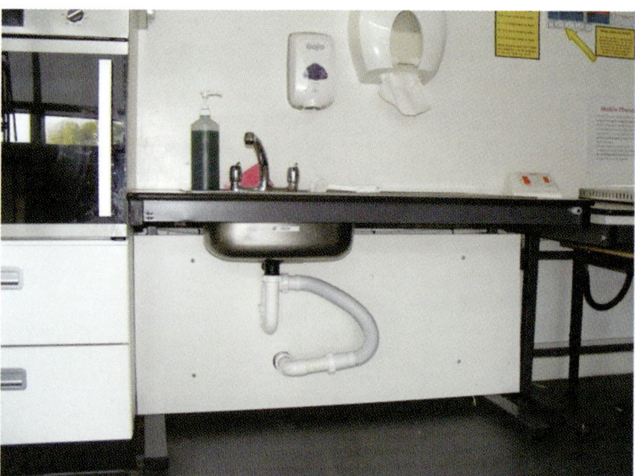

Fig. 5.17 Flexible waste discharge pipe used to connect an adjustable height sink to the drainage connection on the wall.

Fig. 5.18 A whole-house wastewater pump unit undergoing testing. Note the discharge pipe (in glass) with the twin bends before the water flows by gravity into the ventilated branch fitting.

provide the seal to isolate the drainage system from the air. The main appliance with a shallow depth of seal is the bath. Although a bath will produce induced siphonage (hence the gurgling as it discharges) for much of its discharge, at the end there will be a slow flow of water into the trap, to top it up, due to the flat bottom of the bath. This final flow will naturally replenish the depleted trap.

The size, or diameter, of the connecting pipework is mainly determined by the appliance to which it is to be fitted. However, in some situations there will be a choice of size due to the number of appliances connected together. By using larger-diameter pipes and traps airflow will be less restricted, so more appliances can be fitted to a given branch without additional ventilation pipes or fittings. These various factors are brought together in various guidance documents such as BS EN 274 or Building Regulations Approved Document H. This is summarized as follows:

- **WCs with outlets** < 80mm: Trap diameter 75mm, seal depth 50mm
- **WCs with outlets** > 80mm: Trap diameter 100mm, seal depth 50mm
- **Urinals:** Trap diameter 40mm, seal depth 75mm
- **Bath and Shower:** Trap diameter 40mm but can be reduced to not less than 38mm if discharge is di-

rectly into a gully, seal depth 50mm

- **Washbasin and Bidet:** Trap diameter 32mm, Seal depth 75mm but can be reduced to 50mm only on spray tap basins with flush grate wastes without plugs
- **Sinks, food waste disposal units:** Trap diameter 40mm, Seal depth 75mm
- **Washing machines and dishwashers:** Trap diameter 40mm but can be reduced to not less than 38mm if discharge is directly into a gully with grating, Seal depth 75mm.

Most wastewater discharge systems are ridged and fixed in position. This is not a problem in most situations, but where mobile units or appliances with adjustable heights are used, a flexible solution is required (Fig. 5.17). For instance, where appliances are designed to be used by both able-bodied people and those in wheelchairs the use of height-adjustable appliances may be specified.

However, flexible pipework should only be used when there are no better options, it should not be used generally as a substitute for properly formed traps, fittings and bends.

Pumped Drainage Systems

Large-bore pumped discharge units have long been used in locations such as underground railway stations, basements of flats or hotels and anywhere where the existing sewer pipe is above the flood line or the lowest point of a building's drainage system. These tend to have sumps fitted with grinder pumps that can deal with wastewater and reasonable levels of abuse such as the odd rag, plastic toy, or 'disposable' nappy.

The recent trend of adding en-suite facilities and creating 'granny flats' in existing buildings has seen the rise in the application of small-bore pumped discharge units (often called macerators). These devices enable appliances to be located away from drainage stacks, but should only be used as supplementary drainage and not as a sole discharge system. Such devices receive waste from a WC and pass it through rotating blades before pumping the resulting slurry through a pipe around 22mm in diameter. Depending on the pump and unit used, the discharge can be many metres away vertically and/or horizontally.

In some situations, the type of wastewater or location of the pipework may require different materials. For example, in laboratories polyethylene systems may be used for its high chemical resistance along with dilution traps (large-volume sink traps that can contain litres of water and reduce the concentration of small quantities of concentrated corrosive substances).

In some hotels and hospitals, glass pipework is used to ensure the fast identification of any blockages. However, the transparency of a glass system tends to decrease in time, due to deposits and splashing. So unless the system is to be cleaned regularly internally and externally, the visibility of blockage advantage may be lost in time.

All pumped systems are prone to pump failure, blockages in the sump or pipework and electrical failure. The first choice and permanent use of pumped systems are therefore generally discouraged by local authorities and building control. Hence, as far as possible, gravity drainage systems should be used in preference to all others.

Nevertheless, there are a number of products in the markets, for instance, in garage and basement conversions. It is however important to check the compliance and certification of the system before specification.

Chapter Summary

Internal drainage systems comprise discharge branches and stacks fitted with ventilation systems and odour prevention seals. Internal drainage systems do not have to be fitted internally to a building. Most systems operate by gravity, but many now include at least a small section that may be pumped. Pumped systems are useful in basements, below the sewer level, for additional en-suite bathrooms and 'granny flats'.

Although water seal traps have been the accepted norm for odour prevention for centuries, they are prone to evaporation and various forms of siphonage. The development of membrane and other mechanical seal 'traps' now provide viable alternatives to the simple 'U' bend trap.

An alternative to simple open stack pipes, for ventilation, is the air admittance valve (AAV). This only allows air to enter a pipe, but not leave. AAVs are available in a wide range of sizes and applications.

NOTE: [1] Jack, L. B., & Swaffield, J. A. (2009). 'Embedding sustainability in the design of water supply and drainage systems for buildings'. *Renewable Energy,* 34(9), 2061–2066.

Chapter 6

ENHANCED DRAINAGE FOR WASTE-WATER HEAT RECOVERY AND WATER REUSE

Drainage systems cannot be designed without consideration of the water supply system, especially where forms of water reuse or alternatives to a mains supply of wholesome water are being considered. Hence, a clear overview of the intended water supply system must be developed and understood prior to drainage design.

Water can be supplied to a building from the public mains, i.e. through a centralised, municipal supply, or through decentralised, site-based sources. The latter is also known as private water supply in the UK, about which separate regulations apply. A decentralised or alternative water supply can be non-potable (not for drinking) water abstracted from natural water sources such as boreholes, rivers and lakes or rainwater, or from within the building, i.e. water already used for washing and cleaning.

The use of mains water for non-potable uses is increasingly considered wasteful and unsustainable, especially if the energy, carbon and cost associated with centralised processes for mains water supply are considered. So heat recovery from water use or reuse of reclaimed water can help to reduce the climate impact of water processes, conserve limited natural resources, conserve energy and cut in-use costs, i.e. water and energy bills. Non-potable water can potentially replace the following activities:

- Landscape and agricultural irrigation
- Cooling water in building services and industrial applications
- Construction activities, e.g. concrete mixing
- Domestic uses, e.g. toilet flushing.

It is important that a building uses any impinging rainwater as well as any resultant wastewater to the best benefit, thus reducing unnecessary wastage and limiting the loading on sewer and drainage networks and/or collection systems (Jack and Swaffield, 2009[1]). It is therefore worthwhile for designers and engineers to consider and integrate water recycling and reuse early in the design process, or design the supply and drainage system such that it can accommodate these functions later.

This chapter will discuss greywater reuse, i.e. the collection and use of greywater as an alternative to public (mains) or private potable water supply. Greywater is thus diverted from the drainage system for non-potable use such flushing WCs and watering the garden. The separation of different types of wastewater has a number of benefits but can also produce problems. If water reuse is to be carried out, the type of collected wastewater needs to be appropriate for this and any subsequent treatment and resupply of the water for use in the building.

But first, a brief overview of wastewater energy recovery systems.

Wastewater Heat Recovery

There is scope to extract and reuse heat, for example from bathing and washing water for subsequent use for space and hot water heating within the building. Wastewater contains significant amount of heat, which is dissipated as it flows through the sewer to the treatment plants. Discharging heated water from domestic and industrial processes into natural water bodies has been proven to affect the ecological balance of the water, e.g. by limiting the amount of dissolved gases – oxygen and carbon (IV) oxide available to living organisms. This then limits the growth or survival of these organisms.

Energy (heat) recovery systems extract 'waste' heat from bath or shower water and use it to increase the input temperature of 'cold' water feeding water fittings, e.g. taps and shower or water heater, with the aim of reducing the energy needed to raise the temperature of the inlet water. Such devices are now becoming a common installation in new-build dwellings (Figs 6.1 a-b). In addition, systems are available to extract the 'waste' heat from sewers using a network of pipes embedded in the sewer pipe walls. However, this may reduce the temperature of the wastewater in the drainage and sewerage systems such that deposition may become more likely (due to the cooler water) and it may increase the risk of the drains and sewers freezing during the winter months.

In a typical wastewater system, the resources that are more readily extractable and reused include: heat energy, kinetic energy, nutrients and water. Heat energy will originate from appliances such as baths, showers, washing machines and dishwashers. However, developments in white goods, and their surfactants, have resulted in washing machines and dishwashers that operate at lower temperatures, so the potential for heat recovery from such devices may be decreasing. The temperature that people bathe at is normally around blood temperature of 37°C, so this is a more reliable source of heat energy (assuming that people do not sit in cold baths very often). Devices to extract heat from wastewater, and specifically from showers, have been developed and marketed for a number of decades. They mainly consist of a heat exchanger between the outgoing wastewater and the incoming mains-fed cold water, with the subsequent warmed water feeding a water heater, shower or another device that will benefit from using such water. The main devices that can benefit from preheating the water are those with limited energy inputs that produce fixed temperature rises (ΔT) such as instantaneous water heaters or combi-boilers. The energy saved in this way has been shown in UK SAP (Standard Assessment Procedure) calculations for houses to be equivalent to the installation of solar panels, but at a fraction of the installed cost. Hence, they are becoming popular with house builders who want ecological credentials.

Products for capturing heat from domestic wastewater stacks tend to be made from copper or brass (Figs 6.2a–b). Products for shower trays and branches will contain metal parts but are often encased in plastic.

Thus, the wastewater in a drain and sewer system is not simply dirty water. It is heat, momentum, chemicals, heavy solids, floating solids, recyclable material and non-recyclables, all mixed up together in a water carrier. Reducing, or increasing, any one of those elements will have an impact on the drainage performance and design. Work in France has shown that significant reductions in the temperature of wastewater by more than a few degrees can aversely affect the performance of the downstream wastewater facilities (Fig. 6.3). In some ways the ideal transport system for wastewater would be a flexible conduit of variable diameter (somewhat like a human gut), but that would be very difficult to engineer compared with fixed-

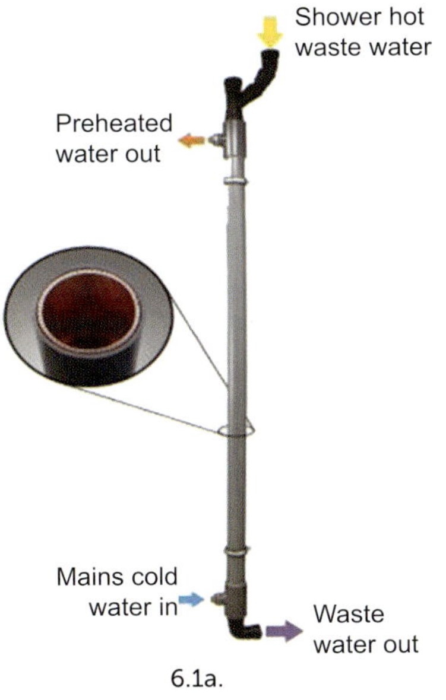

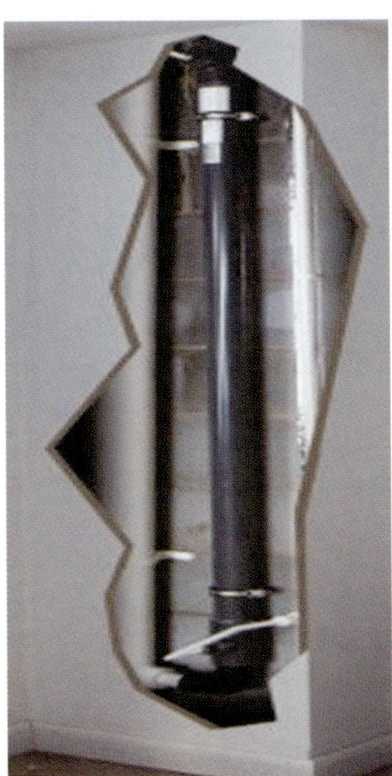

Fig. 6.1 Example shower heat recovery system. (a) Recoup Pipe+ HE Product Diagram with detail. (b) Recoup Pipe+ Heat Exchange Boxed cut-through. Images used with permission.

Fig. 6.2a-b. Sections through a shower heat recovery unit showing the narrow flow paths for the cold water and the drainage cavities between the wastewater and the mains water that are to provide an indication of any failure of the pipe within the unit.

diameter pipes. Therefore, we need to make the best of the materials we have and design systems accordingly.

Greywater Reuse

Greywater systems are categorised based on the extent (quality) to which the greywater is treated and therefore stored. Greywater systems can also be combined with rainwater or integrated sustainable urban drainage systems. Such system types range from direct reuse systems with no treatment, minimal or no storage and with strictly limited applications to chemical, biological, bio-mechanical and hybrid systems.

Some countries have national guidelines for greywater systems including the UK's WRAS Alternative water systems information leaflet and guide, which is free from WRAS, and the BS 8525-1:2010 Greywater systems code of practice for further details. This standard also includes definitions of key terminologies.

Design Considerations

The first step when designing any water recycling and reuse system is identifying the following:

- What is the need and priority for the client and users? This is typically to varying degrees of importance a combination of water, energy and cost savings. However, for some clients, corporate social responsibility, public image, policy, planning or regulatory requirements may be the key drivers.
- What is feasible? What is available within the context, site, building and cost parameters? For instance, what is the annual rainfall for the area? There is no point installing a rainwater harvesting system where there is little rain to service it. Is there adequate collection and storage space within the site, or else is there scope to put storage in a basement or roof? What are the cost implications of these issues?

- What design and performance requirements are specified in local and national guidelines and regulations. There are countries where rainwater harvesting is prohibited for environmental reasons. In addition, there are water quality risks to consider if systems are poorly designed and specified, installed or maintained.
- What is the impact on other functionality aspects of the building? The technical viability of the recycling and reuse systems needs to be balanced against the functional spaces, structure, other service provisions within the building, the site and surrounding areas.
  - For instance, situating a water storage tank has an impact on the structural design.
  - Also, pumps, monitoring and control systems are needed to support the system and this will require allocated plant room space.
  - Decisions should be made about where and how wastewater is collected and transmitted for storage, retreatment and redistribution.
  - The impact of diverting wastewater from drainage systems needs to be considered as reduced flow capacity may negatively affect drainage systems if not considered in the initial design and implementation strategy.

The last point is particularly important with the increased use of water-efficient fittings in buildings, e.g. low-flush WCs, and eco-taps and showers. Problems in existing drainage networks were documented in some parts of the USA where, unlike the UK, the WC flush volume was gradually reduced over about four decades from 9 to 6l, many states implemented a quicker switch from 3gal WCs to 6l ultra-low flush WCs. The increase of blockages, due to lower 'drain-line carry' distances, angered many. However, the fact that many of the problematic drains had been faulty before the change in WCs was not readily acknowledged. To date, some regulatory bodies and some pro-

Fig. 6.3 CSTB water research laboratory in Nantes, France, with the heat from the wastewater test facility in the foreground.

fessionals have used this as an excuse not to innovate and deliver water-sustainable buildings by avoiding specifying for these through either alternative water supply or the use of new water-efficient fittings.

With good collaborative design, and by engaging all the stakeholders including the water companies during the early design decision-making stages, it is quite possible to have a water-efficient building that has little or no impact on the drainage system. There are also many technological solutions that can be specified to reduce risks. Examples include devices that store wastewater until there is sufficient volume to discharge a strong flush wave through the system. These are basically tipping tanks of various designs. Some have no moving parts and use syphonic action to discharge, while others simply rely on gravity. Therefore, when designing a drainage system with high levels of reuse and water efficiency, a drain flushing discharge unit may be a desirable, if not an essential, addition.

Typical design specifications for the capture of rainwater and greywater from drainage systems within and external to the building include:

- Anticipated available flow rates from wastewater sources, e.g. baths and showers.
- The range of static and dynamic pressures that are available through a typical twenty-four-hour period.
- Anti-surcharge and backflow prevention requirements. Surcharge and backflows can occur during flooding and it is critical to ensure that no contamination occurs during such events.
- Temperature range of the wastewater.
- The type and bio-chemical nature of the wastewater – for instance, foul water from toilets is typically not collected for reuse. Neither is water from kitchen and food preparation areas due to high concentrations of grease, detergents, soap and vegetable matter. Similarly, stacks with only baths and showers can also suffer from detergent-related problems. It should, however, be pointed out that developments with detergents over recent years have reduced foaming – and the use of low-temperature washes has meant that detergents now tend to be soluble for longer in the wastewater.
- Position, type, size, capacity and layout of the pipework. The nature of the wastewater will also influence the material and gradient of the pipework.
- Position, type and size of connections.
- Plant location and sizes, including capacities for hot and cold water systems (storage, pumps, accumulators, etc.), hot and cold water storage, treatment and pressurisation for supplies, waste storage, wastewater and stormwater storage, possibly wastewater treatment and collection.
- Volumes of recycled water demand to inform percentage collected and stored versus percentage of the wastewater discharged and removed from the building via the mains sewer.
- Determine main pipe and drain routes from floors to and from risers. Include other sources, e.g. rainwater harvesting.
- Establish main below-ground drainage routes and access locations for inspection and maintenance.
- Coordinate above and below-ground drainage with the rest of the design: superstructure (walls/floors) and substructure (foundations).

One way to determine the potential for water reuse systems is to monitor the discharge of appliances within a representative building. This can be done by using individual meters on supplies, but it involves disruption to the plumbing systems, the fitting of water meters and setting up a data recording and analysing system. Although this is reliable, it is very expensive to conduct and the work in removing the meters after such a trial can be even more disruptive and introduce increased possibilities of future leaks. An alternative is to use an intelligent meter that can determine the appliances that are being used from their characteristic flow rates. Such devices are often provided on a rental basis by the suppliers, who will analyse the data as part of the rental.

Such devices tend to be very expensive to buy and take time to set-up to 'learn' the various characteristics. However, they have now become trusted apparatus

Fig. 6.4 Backflow device.

that is often used by water companies in large-scale water efficiency trials. Recently, more compact units have come to market that are intended for permanent installation, not temporary. These new meters are linked via the data cloud to analysis companies that monitor the data for the building users and can provide fast warnings of leaking pipes or appliances as well as consumption data from individual appliances, using the established characteristic flow patterns from the supplies.

All of these characteristic-driven meters are good when there are single uses, but with multiple simultaneous use they can lose data when the individual characteristics become undiscernible. With sophisticated software the amount of 'lost' data can be minimised to the extent that the analysed data is usually more than accurate enough to inform decisions on water consumption and the potential for savings and reuse. Such meters will become more widely used in the future as smart drainage systems become established and integrated water system design and operation will need such data to perform sensibly and reliably.

The last step is to produce architectural drawings, details and specifications to reflect the best option possible.

Greywater Collection

Greywater reuse should be from collected wastewater,

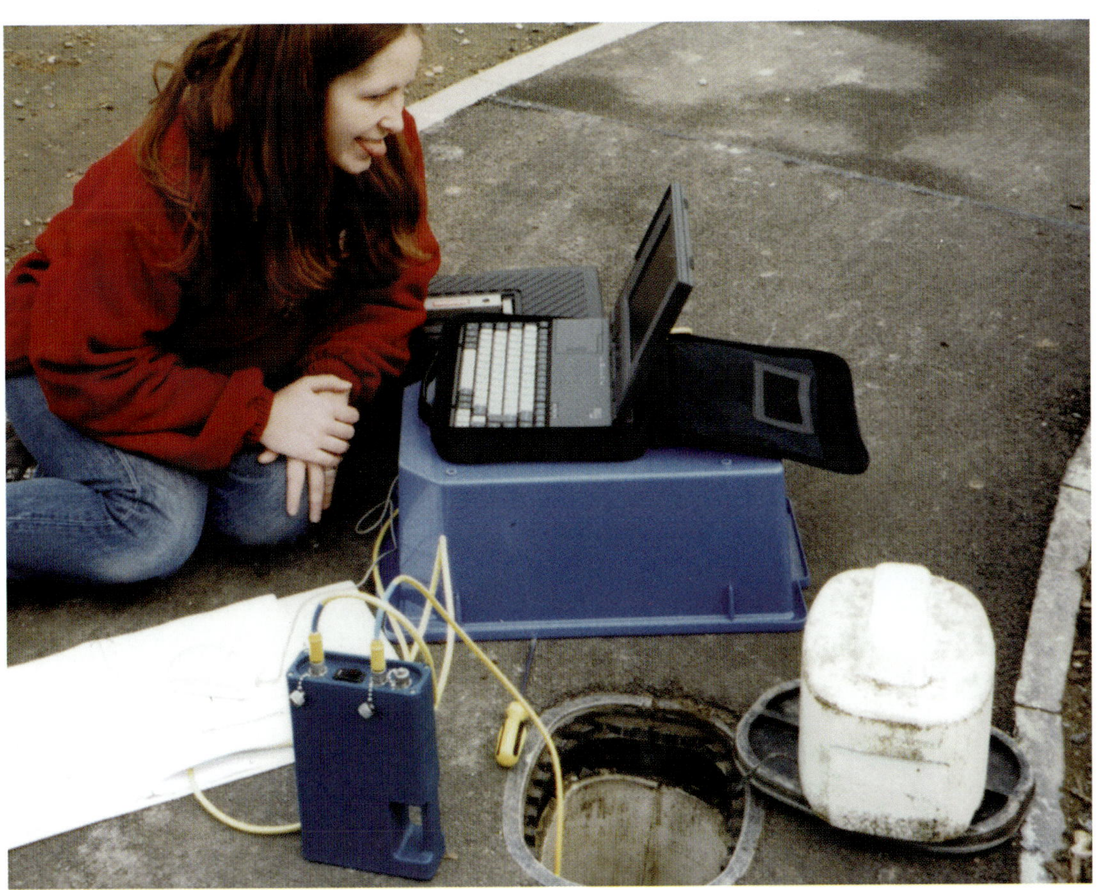

Fig. 6.5 Data being analyzed from a 'trace wizard' meter attached to the water supply feed to a house.

not foul or blackwater, e.g. from toilets and urinals. Therefore, only wastewater from bathroom sources, e.g. baths and showers, is collected typically. Poorly designed rainwater and greywater systems can have a negative impact on the drainage systems and increase the risk of contamination of the water supply. Thus, wastewater should be collected via an enhanced or separate drainage pipework to avoid compromising the normal functioning of other drainage systems in the building.

A bypass device should also be installed so that the wastewater can flow directly to the sewer if the system develops problems, or is undergoing repair and maintenance. The pipework size and layout should follow the respective guidelines, e.g. BS EN 12056 Part 2 (2000) to minimise turbulence and foam from soaps. This includes minimising the volume of foam that is discharged into the foul drain and sewer system. Accessible and visible traps and filters should be specified and installed, preferably at source, to minimise blockages in the pipes. This also means that bends should be avoided as much as is feasible and the wastewater should be free draining to avoid stagnation. This means that provisions should be made for the complete emptying of the complete greywater system, including all the pipework, tanks, filtration and treatment units, and pumps, either by accident, as a result of periodic maintenance or a process of decommissioning the system.

Lastly, all water recycling and reuse cisterns and tanks should be closed, water tight and vented. All storage tanks and cisterns should also be marked and labelled adequately so that no mistake is made regarding their contents.

Greywater Distribution

After collection, filtration and/or treatment, treated greywater for reuse is then redistributed for use within the building, e.g. for flushing toilets and urinals, or external to the building, e.g. gardening and irrigation. The waste or treated greywater can be collected or distributed by gravity, or by using pumps.

Fig. 6.7 shows a typical greywater process with inlet pipes bringing in wastewater via gravity or pumps from bathroom sources, and outlet pipes either delivering the treated water via gravity or pumps directly to the point of use, or to an intermediate tank close to the point of use.

Pipework materials can include different types of plastics, copper and stainless steel as long as they are sized to provide adequate flow and cope with the pressure generated within the system. Note that oversized pipes may lead to low flow and undersized pipes may result in high pressure, burst pipes and leakage. Multi-

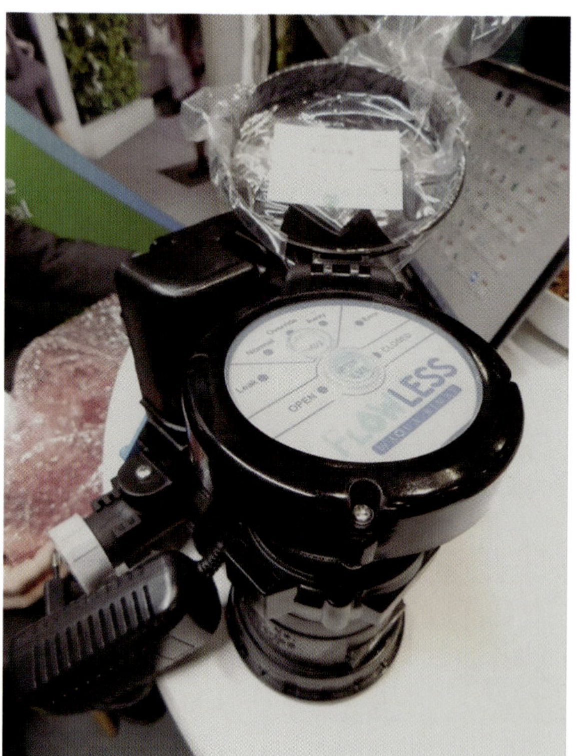

Fig. 6.6 A water intelligence unit that can provide appliance use data to the householder, landlord or others as well as provide warnings about leaks or faulty appliances.

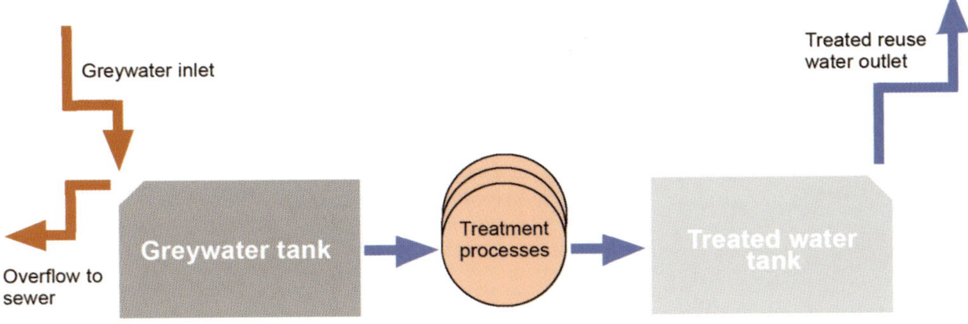

Fig. 6.7a Typical greywater process.

layered pipes may also be used, typically to protect against temperature fluctuations as in the case of insulated pipes, to preserve the heat for later capture or to recover the heat in situ for reuse elsewhere.

The pipework connection from greywater system outlets should follow the steps and guidelines provided in the previous chapter and in the relevant national and international regulations that inform the choice of pipe materials, sizing, layout and other requirements to avoid risks to the building and its users.

Pipework used to collect wastewater or distribute treated greywater must be identifiable from other water sources in the building throughout its lifetime. All internal pipework for the captured wastewater and water supplied for reuse must be colour-coded

Figs 6.7b and c Wastewater inlet and greywater outlet tanks.

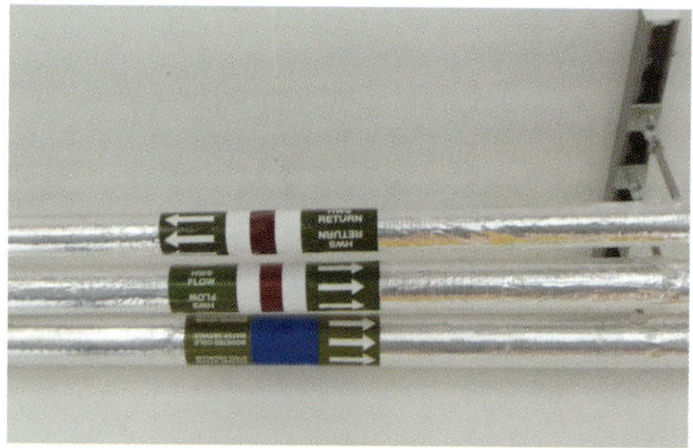

Figs 6.8a–c Labelled pipework outside a toilet. It should be possible to deduce the content and direction of flow of what is being conveyed in the pipework from the label. Note the white colour in this instance is to achieve sufficient contrast to make sure that the safety colours are easily read.

6.8a

6.8b Example of service identification of drinking water from public water supply (light grey represents the natural pipework colour).

6.8c Example of service identification of non-potable water from greywater systems (light grey represents the natural pipework colour).

Pipe contents	Basic identification colour	Safety and code colours	Basic identification colour
Pipe diameters up to 50mm	50mm min	30mm	50mm min
Pipe diameters 50 – 100mm	100mm min	75mm	100mm min
Pipe diameters > 100mm	150mm min	150mm	150mm min
Source of water			
Potable water derived from the public, mains water supply	Green	Auxiliary blue	Green
Potable water derived from any other source, e.g. borehole	Green	Flint Grey	Green
Water quality			
Non-potable water system derived from any other source	Green	Flint Grey / Black / Flint Grey	Green
Non-potable water system derived from public water supply	Green	Flint Grey / Black / Flint Grey	Green

Table 6.1 Example colour stripes for water services. (Adapted from Figure F1 and Table 2, BS 1710:2014.)

and labelled appropriately (Fig. 6.8).

Guidance on identification and labelling in the UK can be found on the WRAS (UK's Water Regulations Advisory Scheme) Pipe Identification Information Note. This summarises the requirements of the EN 1710 (2014): Specification for identification of pipelines and services. The explanation for the use of the colours are shown in Table 6.1. Clearly labelled valves and outlets are also essential. This is again to avoid cross-contamination with the mains water supply. Pipes buried below ground may need to be identified differently to those above ground due to existing conventions for services in and around buildings.

Identification: Mechanically secured tags or self-adhesive labels should be added to pipework bearing wastewater for reuse or redistributing treated greywater to appliances and fittings. Tags identifying the water supply to each appliance should also be added at key connection points using flexible fasteners. The tags and labels should not be less than 100mm in length, coloured or edged in GREEN and have 'GREYWATER' and any additional information in black lettering not less than 5mm in height. Avoid using identification codes alone.

Lastly, simple, easy to understand signage should be provided at any point of use, including all appliances. The signage should provide supporting text and tactile signs, e.g. braille, and specify the type of water if other non-potable water is provided in the building.

Control system: All systems should have a control unit. An automatic control unit should ideally be specified and installed that monitors the effective operation of the system and then automatically shuts it down and diverts wastewater to the sewer if there is a problem. It should include visual and audible warnings and alert system owners to operation and maintenance protocols. It should activate the back-up mains water supply and trigger the bypass of wastewater to the foul water sewer if and when necessary. The control system

should also manage the supply and demand of, and volume and duration of storage of, untreated greywater beyond the period specified in water quality standards as well as the system manufacturer's specifications. The control unit may be connected to the overall building management system (BMS) to aid the holistic management of the building's resource use.

Backflow Prevention

The main risk of greywater systems are contamination due to poor information about the water source, dangerous or indiscriminate pipe connections, backflows and surcharges. Dye testing of the pipework is therefore conducted before the mains or back-up potable water system is commissioned. Meanwhile, the system

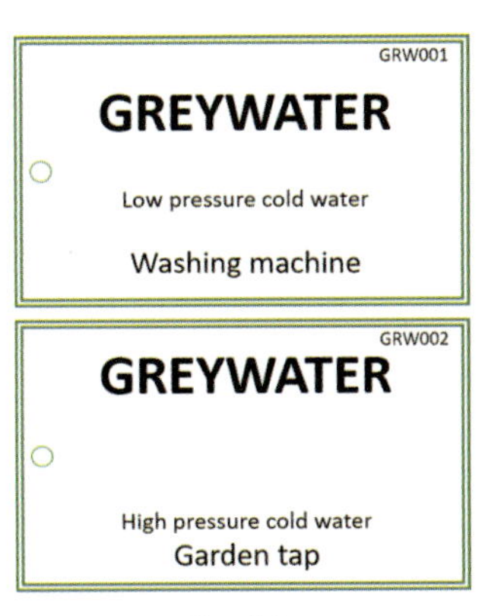

Fig. 6.9a

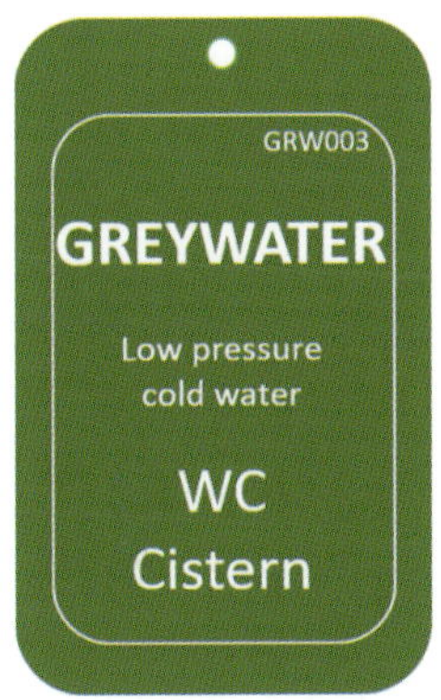

Figs 6.9b

Figs 6.9a-b Labels for pipework and at key connections.

Fig. 6.9c Labels at point of use.

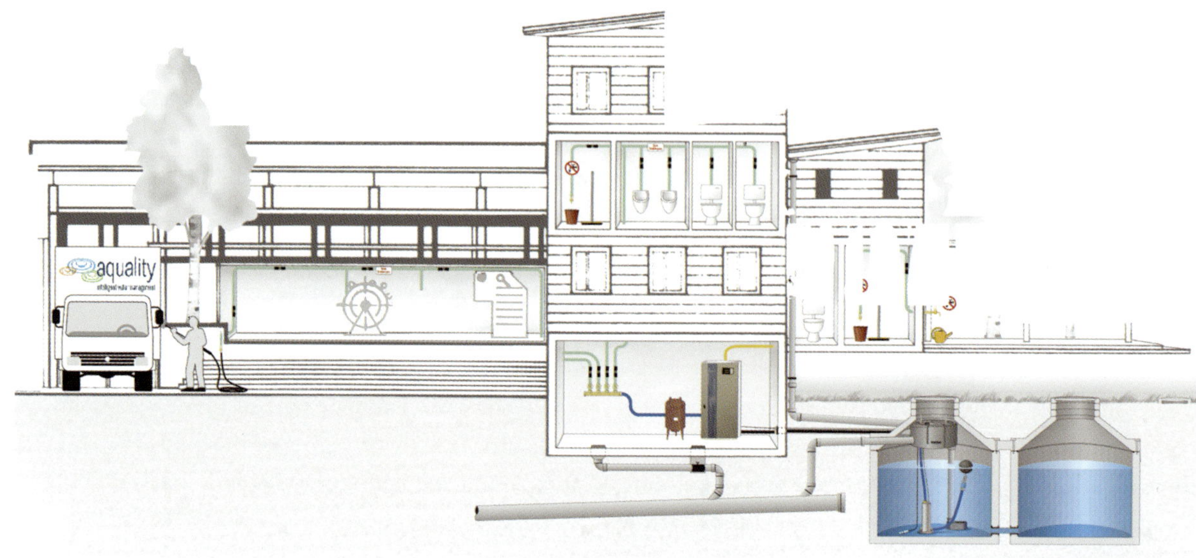

Fig. 6.10 Schematic of a greywater reuse system with an immersion pump and showing colour-coded pipes and labels at point of use. (Source: Aquality. Used with permission)

should include:

Air gaps: This is an important plumbing principle that is worth remembering and explains why the outlets of taps do not align directly with the highest fill points of basins and sinks. In essence, the air gaps serve as a safeguard for preventing the backflow of grey non-potable water into potable water, thereby causing contamination.

The UK's Water Fittings Regulations specifies the need for an air gap in any recycling and reuse system. These requirements are supported in the BS 8525 (2010) Greywater Systems, which specifies two types of air gaps – Types AA (unrestricted discharge) and Types AB (weir overflow) in compliance to BS EN 13076. Devices to prevent pollution by backflow of potable water. The principle to remember is that there must be a clear, visible and unobstructed physical air break (gap) between the lowest level of water discharge and the level of potentially contaminated water down-stream of the system. Alternatively, this should be at any point where different types of water or water supplies meet. In addition, there must be no direct pipework connection to bypass any airgap or backflow prevention arrangement or device. The discharge or connection could be to a cistern, tank, fitting or within an appliance. For compact installations where only low flow rates are required, the type AD water injector air-gap device can be used. AD air gaps can be built into appliances and pipework.

In most cases, a simple tundish device, as shown in Fig. 6.12b, can be used.

Anti-surcharge and backflow prevention device: Surcharge can occur during flooding within or outside the building, so in addition to air gaps it is important to install an anti-surcharge device to prevent backflow in pipes and appliances. The device should conform to BS EN 13564 Part 1.

Break cistern: Waste or non-potable water of any kind

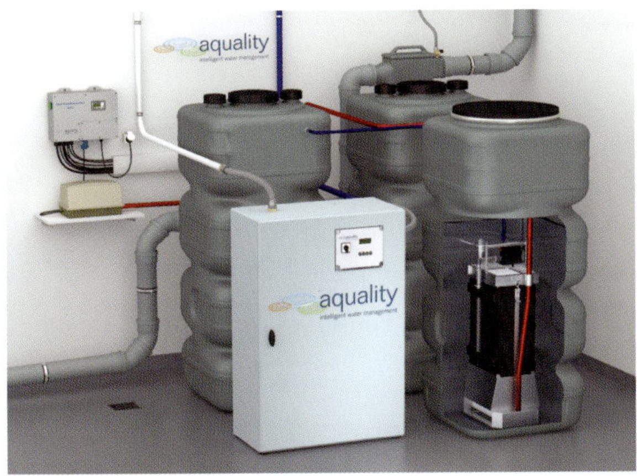

Fig. 6.11 Compact greywater system and control panel.
(Source: Aquality, Used with Permission)

Fig. 6.12a.

should never be connected directly to a mains or potable water supply. Therefore, a cistern or tank should be specified to physically separate the two drainage and supply systems. This typically means a cistern to collect the greywater from the drainage system and a separate cistern to store and resupply the treated water for reuse.

Calmed inlet: The end of the drainage pipe feeding the cistern or storage tank should have a fitting that slows the water flow into the tank and minimises turbulence within the tank.

Non-Gravity or Pumped Systems

It is feasible, and makes economic sense, that the wastewater is collected by gravity. However, it may not always be feasible to distribute the treated greywater via the same means. Therefore, pumps are used to ensure that the treated water is readily available at the point of demand and at the right pressure.

BS EN 60335-2-41:2003+A2:2010, 'Household and similar electrical appliances. Safety. Particular requirements for pumps' provides guidance for pump specifications. Most greywater system providers also typically supply pumps as part of a complete provision.

Fig. 6.12b.

Figs 6.12 a-b shows (a) minimum airgap in a sanitary appliance (b) in-line tundish for air gap in pipes

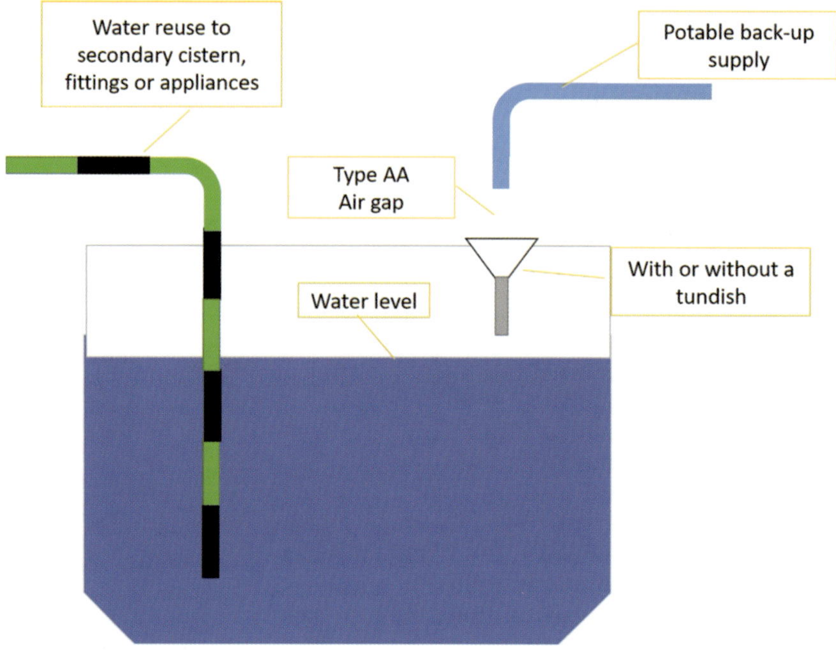

Fig. 6.12c

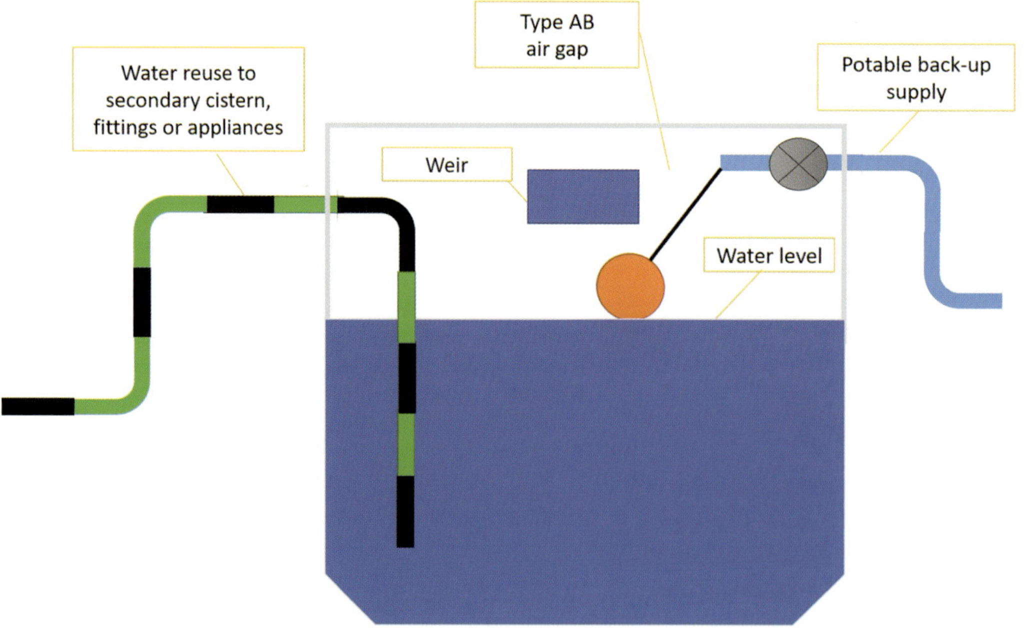

Figs 6.12c-d Air gap for cisterns.

It is, however, important to check the constituent parts including the pumps for compliance, or seek warranties of the system as a whole when purchasing.

The key thing to check is that the specified pump is of sufficient power to overcome the static lift and friction losses in the pipework and valves in order to deliver the water to the point of use. Power specifications for a two-storey building will be different to that of a seven-storey building. Therefore, decisions about whether to use single or multiple pumps, pressure control and regulating valves will need to be made, while considering energy use and noise. The pumps should be ideal for the anticipated level of demand. Also, air intake, cavitation and water hammering should be prevented as these affect the functioning of pipes. Wastewater usually contains particulates and contaminants that should be removed or filtered as they may result in blockage or affect the normal functioning of the pumps.

Pumps can be immersed within tanks or installed outside them. For the former, a minimum level of water must be maintained about the pump inlet to prevent it sucking in air or debris at the bottom of the tank. If installed outside the tank, the pump should be ventilated, self-priming and the suction line to the pump should have an upward gradient to the pump. The pump mounting should be sound and vibration proof.

Beyond these, the guidelines pertaining to gravity systems also apply, including the need for filters, access for maintenance, back-up and backflow provisions including isolating and non-return valves as well as system control. Pumped systems are also required to have control units, which function in a similar manner to gravity-based systems, as well as:

Ensuring there is a manual override for pumps in case of faults, e.g. pumps running dry, electric overloading or overheating.

Preventing the pump from running indiscrimi-nately and perpetually. Pumps should always run according to demand.

Integrated Greywater, Rainwater and SuDS System

In certain instances, it may be beneficial to combine greywater, rainwater and to some extent surface water attenuation schemes in order to meet the non-potable water demand, especially in large development schemes or as a site-wide, neighbourhood level solution. In this instance it is important to investigate the feasibility of such an approach in the context of the site, design and planning constraints, location (e.g. amount of annual rainfall, flood risks), cost benefit for the client, etc.

It may be possible to design pipework that combines all the water sources into a single storage prior to or after treatment, or to design each system separately. These systems can operate independently if different levels of water quality and treatment is needed, if systems from different suppliers are being used, or if each water source is going to be collected and used in different parts of the building and site.

If systems are integrated before treatment, it is important to consider volume control and management. Considerations should include: the different types of water; the type and nature of required treatment for each source; how each water is going to be collected and redistributed for reuse, e.g. gravity versus pumped, energy use; and if energy be captured and reused in the system, e.g. for pumping, mechanisms to manage the system if under or over capacity, etc.

If systems are integrated after treatment then the guidelines pertaining to each system should be followed. For example, EN 16941-1:2018 'On-site non-potable water systems. Systems for the use of rainwater' for rainwater recycling, and SuDS and BS 8525 for greywater systems.

Water reuse and energy recovery are new areas for drainage design and are vital in changing the concept from one of disposal of waste to utilisation of resources. However, there are limitations to reuse and energy recovery.

Each time water is reused the concentration of pollutants can be increased, or decreased, depending upon the treatment system that is used. If more concentrated effluent is produced it can have a detrimental impact downstream at the wastewater treatment plant.

Similarly, if too much energy is extracted from wastewater it will be more at risk of freezing and the wastewater temperature can be too low to enable the wastewater treatment plants to work as intended.

There are now many water reuse systems available. Generally, the larger the system, the higher the viability. Long-term maintenance needs to be ensured with all systems.

Where water for reuse is available, the possibility of a cross-connection – resulting in contamination of drinking water – needs to be prevented. Clear marking of pipes and fittings is needed.

NOTE [1] Jack, L. B., & Swaffield, J. A. (2009). 'Embedding sustainability in the design of water supply and drainage systems for buildings'. *Renewable Energy,* 34(9), 2061-2066.

Chapter 7

DRAINAGE AT WATER-USING APPLIANCES AND FITTINGS

ater efficiency is achieved through the 3R principles of reduce, reuse and recycle. The design and specification of water products, fittings and fixtures in a building helps to reduce water consumption at point of use. In the UK, sanitary appliances in new buildings are required to meet the baseline standards for water use as shown in Table 7.1. However, designers and building providers are encouraged to exceed this minimum requirement.

Appropriately designed water fittings such as pipes and traps ensure that water is supplied efficiently to appliances and outlet fittings, then wastewater is collected and transmitted effectively from them for reuse and recycling, or for treatment and disposal.

It may be tempting to reduce drainpipe diameters where water efficient measures are employed but this increases the risk of drain blockage. The premise is that the larger the pipe, the lesser the risk of blockage. This is why many sewage undertakers and wastewater authorities insist on 150mm diameter drains as a mini-

Fittings/Appliances	Recommended Baseline	High Efficiency Options	Performance Features
WC	6l/flush	$\leq$ 3.5l/flush	Dual flush, pressure assisted, power/air assisted toilets, vacuum, composting
Shower	8l/min	$\leq$ 6l/min	Aerating, laminar flow showerheads or flow regulators
Tap (Basin)	Up to 12l/min	$\leq$ 4l/min	Spray tap, sensor-activated tap, push tap or tap flow restrictors
Tap (Kitchen)	12l/min	$\leq$ 6l/min	
Urinal	3.8l/bowl/flush	$\leq$ 1.5l/bowl/flush	Hydraulic valve, passive infrared sensor, timed flush, waterless urinals
Bath	200l capacity (excluding body mass)*	$\leq$ 155l capacity (excluding body mass)	(Bath performance features) Contoured, raised base
Washing Machine	10l/kg dry load**	$\leq$ 7l/kg dry load	Load sensing and dirt sensing
Dish-Washer	1.2 /place setting***	$\leq$ 0.7l/place setting	Load sensing and dirt sensing

*(The UK regulations specify maximum 230l capacity. A higher capacity requires notification)

**(The UK water regulations specify maximum consumption of 27l/kg for wash load for a 60c cotton wash)

***(The UK regulations for domestic dishwasher is given as 4.5l/place setting)

Table 7.1 Baseline figures for appliances and fittings.

mum, in order for them to provide any service to the connection. Sanitary fittings outlet pipes also have specified minimum internal dimensions, as will be discussed in this chapter. Hence, any reduction in the volume of water from a building will need to be taken into account so that drain and sewer performance is not compromised.

Therefore, water efficiency may impact on drain and sewer systems. It is important to note that this is because most legacy sewer systems operate based on gravity and require certain volumes of flow to operate properly. New sewer systems are designed to cope better with reduced flow.

Still, too much or too little water in old sewer networks may cause problems. If a large development of water-efficient buildings is planned, special measures may be needed to maintain drain and sewer flows. For example, 'tipping tanks', which have been developed to provide periodic flushes of wastewater to ensure cleansing of the drain and sewer pipes. Although some are simply containers that when full tip over and empty their contents, there are many that contain no moving parts and utilize syphonic action or air compression to provide a pre-determined volume of flush water to the pipes.

Therefore, designers should consult the right professionals including the water companies during the early stages of design, so that the right calculations are made based on the coping capacity of the existing sewer network and that there is a robust understanding of the inputs and outputs of drainage systems in any development. It is also necessary to understand the impact of the drainage systems on a development on a neighbourhood, town or city network to ensure that new and existing sewers function optimal to their intended purpose.

Drainage at Appliances and Fittings

The discussion up to this point has focused primarily on pipework design, components and specifications, as well as connections of pipework to separate or combined sewers.

Drainage at appliances and fittings is generally concerned with the following:

- **The waste outlet,** i.e. the junction or fitting that connects the appliance, such as a washing machine, or fitting, for example the kitchen sink, with the pipework for conveying the waste away from the point of generation. This junction should be sealed appropriately to prevent leakage while maintaining access for inspection, repair and maintenance. In the case of washing machines or dishwashers, air gaps are required to prevent siphonage. This is often provided by a looped discharge hose. All waste outlets in sinks and baths should be fitted with a fixed or removable plug and grille. The latter is to minimize the amount of solid waste objects, such as food waste and hairs, being discharged into the pipes and drains. This helps to reduce blockages. Plugs are required for baths, basins and sinks unless they are: fitted with spray taps, self-closing taps, or where water is delivered below 0.06 litres per second.

- **Backflow management.** As discussed in previous chapters, this is a hydraulic or mechanical seal between the waste outlet and the pipework to prevent foul air returning into the building via the appliance or fitting. This is typically achieved with the use an appropriate 'trap', which does not obstruct the discharge flow in the manner in which it operates. All traps should be specified and installed in such a way that they can easily be cleaned.

- **Overflow management.** Overflow devices should be fitted to any sanitary appliance to ensure that excess water or wastewater filling up to a certain level

Dimension	Values (mm)	Remarks
Depth of water seal	≥ 50	For partly filled discharge branches in accordance with EN 12056-2 (type I and II)
	≥ 75	For full bore discharge branches in accordance with EN 12056-2 (type III)
Connection to the discharge pipe	ISO 228-1 G1 1/4 B	Wash basin, bidet
	ISO 228-1 G1 1/2 B	Kitchen sink, bath, shower tray
	ISO 228-1 G 2B	Kitchen sink
	DN/ID 30, 40, 50, 60	In accordance with EN 476:2011
	DN/OD 32, 40, 50, 63	In accordance with EN 476:2011
Length of straight part of plain ended trap outlet	≥ 30	--
Length for connection through walls	≥ 245	--
Thread of nuts	ISO 228-1 G1 1/4	Wash basin, bidet
	ISO 228-1 G1 1/2	Bath, kitchen sinks, shower tray
	ISO 228-1G 2	Kitchen sink
Useful length of thread	6,5–10	Nut made of metallic materials
	8–11	Nut made of plastic materials
Total height of waste outlet with trap	≤ 83	Shower tray
	≤ 128	Bath, shower tray with overflow; Shower tray with vertical outlet
Useful length of thread of waste outlet	≥ 11 [a]	--
a) The first full diameter thread shall start with 2mm from the end of spigot.		

Table 7.2 Dimensions for sanitary appliance traps. (Source: Table 2, BS EN 274-1).

in the appliance is still discharged properly. Overflow connections are typically made to the waste outlet and not the trap. No overflow connection should be made at any point of the outlet part of any type of trap.

- The purpose of a good drainage system is to ensure the flow of water and wastewater along the correct channels through the building, the site and built environment. Indiscriminate flow of wastewater through a building will be undesirable to users, cause considerable damage to other building materials, fixtures and fittings, and could cause consequential health risks, e.g. through fungal and microbial growth.

Although appliances such as basins, sinks, baths and shower trays may be fitted with overflows, they are only intended to accommodate the flows that would be expected from a leaking tap; not full bore flow from all taps at once. This means that if an overflow occurs, the water may eventually spill over the appliance and cause significant damage to the building, with risks to the users. One solution to this issue is to fit a supply shut-off device that is overflow activated (Figs 7.3a&b). Such devices have proven to be reliable, but can be quite large.

It is worth noting that most sanitary appliances and products in the market come with waste outlets, traps and/or grille and overflow devices. These either arrive as one combined unit or as separate, to be mechanically fixed (e.g. threaded and screwed, push-fit or clipped together) components. BS EN 274-1 is a useful reference for the requirements for waste fittings in sanitary appliances.

Sanitary appliances should be selected based on:

- The amount of space available
- Required functionality, e.g. durability, ease of cleaning, accessibility
- Types of users and typical duration of use
- Resource consumption – water and energy associated with water use (e.g. hot water), and running costs including costs of maintenance.

See BS 6465-3: Code of practice for the selection, installation and maintenance of sanitary and associated appliances for details. The next sections will give an overview of other specific requirements and design considerations for the main appliances and fittings in a building.

Kitchen Sink

The kitchen sink is used to aid food preparation, cleaning food and food preparation utensils and generating this and other associated wastewater as part of the functional activities in a kitchen. Kitchen sinks can be wall hung, inset, mounted, under-mounted, etc., according to design or user choice, as long as the pre-

discussed general guidelines involving outlet, backflow and overflow management are followed. Also, all areas surrounding the sink drain effectively into the bowl and the outlet at the bottom of the bowl. The BS EN 13310:2015+A1:2018 provides detailed guidance on the functional requirements and test methods for kitchen sinks. This includes guidelines on the material and performance requirements and how these can be tested. In commercial kitchens, floor and channel drains are common. Tube plugs are often used to provide large or internal overflow.

The kitchen sink, as with all sanitary fittings except the WC, is not designed for channelling solid waste. Therefore, it is important to ensure that the appropriate type of waste outlet is used to minimize the inflow of solid waste into the drain network. However, modern kitchen sinks now come with one of two types of waste outlets as standard: (a) a small brass or chrome outlet hole with a rubber plug. Strainers may have to be purchased separately, or (b) a larger outlet with a strainer or basket plug (Fig. 7.1). This is more common in modern sinks and they now come in a remarkable array of different colours, materials, types and configuration to suit user taste and requirements.

Most kitchen sinks come with a 90mm sink outlet

Fig. 7.1 A strainer plug, rubber plug outlet with overflow, combination waste, trap and appliance connection spigot.

Fig. 7.2a Double-bowl sinks. Sink outlet with integral overflow on main bowl only. Also showing spare spigot.

hole or gap, but they can also come in other sizes. The 90mm diameter outlets enable the optimum installation of food waste disposal units. It is therefore important to make sure that the waste outlets, overflows, traps and flanges are appropriately sized and installed to suit the sink material and the size and type of the sink outlet. These useful sizes and dimensions are summarized in Table 7.3 below.

Washbasins

Washbasins are typically used for upper body cleaning and washing, as well as any other type of non-food washing. They are typically situated in public toilets, cloakrooms (in the UK, this is a common term for a room with toilet facilities), shower rooms and bathrooms in most types of buildings. Wash basins can be with or without tap holes. They can also be wall hung, bracket or pedestal mounted. The types of traps also vary depending on the type of washbasin and size of

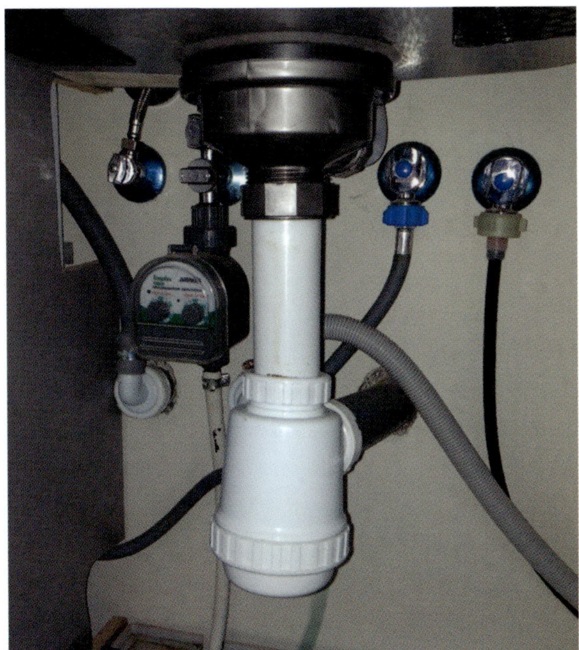

Figs 7.2b-c Waste outlet connection for the half-bowl of the kitchen sink. Note that only the main bowl has an overflow and each outlet has a separate trap. Also visible is the spigot for connecting the waste pipe from the dishwasher on the other side.

Figs 7.3a-b A sink with and without an overflow activated supply shut-off device.

waste outlets. They come in a wide variety of materials, shapes and sizes too, and can even be washing troughs rather than actual basins. See BS EN 14296:2015+A1:2018 for guidelines for communal washing troughs.

Washbasins also need to comply with the general guidance for sanitary appliances and have correctly sized and connected waste outlets, draining, and back-flow and overflow management. It is worth noting that washbasins, especially in non-domestic buildings, can take many forms. The BS EN 14688:2015+A1:2018 provides guidelines on the functional requirements and test methods for washbasins.

The waste grills of washbasins are slightly different to those of kitchen sinks (Fig. 7.4). This is because there is typically less need to filter solids.

Waste fittings with integral plugs tend to have minimal obstructions inside. The restricted inlet to the fitting helps in preventing foreign bodies entering the drainage system. The pop-up waste fittings often have a grating below the plug in place of the fixed gratings in wastes for fully removable plugs. The gratings have many forms, such as a cross or a perforated disc, but they are to stop items such as rings falling into the waste and minimize the amount of hair that might be washed away. Hair mixed with toothpaste and soap will form robust deposits in the pipework, so minimizing their entry is wise.

Fig. 7.4 Types of wash basin plugs including the pop-up and flip plugs. Note the white plastic grating on the pop-up plug's tail.

Figs 7.5a-b Basin with pop-up plug. Note the plug operating rod behind the tap that connects to the small lever inside the waste fitting, above the white plastic discharge pipe. Note also the slots in the waste that drain the water from the overflow.

Figs 7.6a-c Different types of waste and gratings: plugged and unplugged basin, sink with basket strainer plug.

Figs 7.7a–b. (a) Waste outlet to a ceramic washbasin. The metal rod operates the plug. (b) Bottle traps are commonly used in washbasins even if hidden behind a pedestal mount.

Figs 7.8a–b. (a) Typical single tap ceramic wash basin with overflow moulded as part of the basin. (b) Wall hung bidets with tap supply and hidden waste.

Pedestal-mounted washbasins in a test laboratory.

Fig. 7.8c Multisink basin assembly with hidden wastes, taps and pipes behind a removable panel.

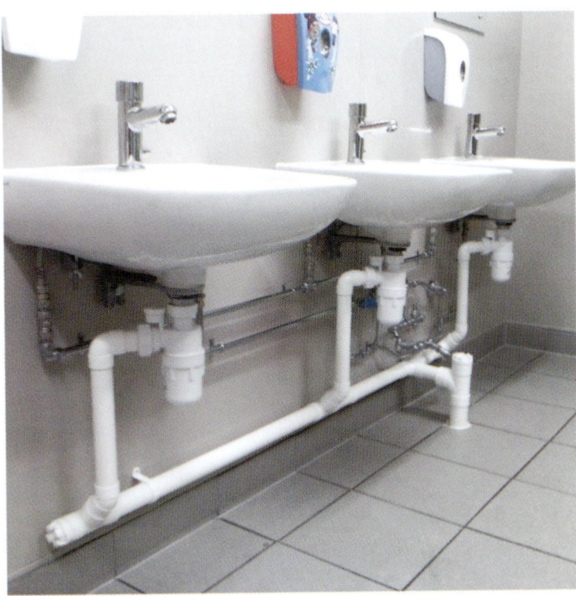

Fig. 7.8d Basin installation with exposed pipes and featuring anti-vacuum traps with miniature air-admittance valves.

It is important that washbasin waste outlets, over-flows, traps and flanges are appropriately sized and in-stalled to suit the sink material and the size and type of the sink outlet. These useful sizes and dimensions are summarized in Table 7.4.

Baths

Baths or bathtubs are sanitary fittings used for wash-ing the body either by full or partial immersion. They also sometimes include a shower installation at one end of the bath. Baths (and bathrooms) are typically prevalent in domestic buildings, whereas showers have more versatility of use in other buildings includ-ing workplaces, sports and leisure facilities, hotels, etc.

The BS EN 14516:2015+A1:2018 guidelines for baths for domestic purposes includes recommenda-tions for the material as well as the type of loading support of the bath. The general guidelines for waste outlets, backflow and overflow also applies. The latter is more important as the overflow of baths can cause

significant damage to the building, in addition to caus-ing health and safety risks to its occupants.

All baths should have at least one waste outlet. All water entering the bath, with the exception of that re-tained by surface tension, should drain efficiently via this waste outlet. The trap should provide access for cleaning out hair and other items that may cause blockage to the flow.

Useful dimensions for the waste outlets for baths are summarized in Table 7.5. More detailed informa-tion can be found in BS EN 232 (2012), Connecting dimensions for baths.

Showers

Showering is an increasingly popular option for wash-ing the body than taking a bath. In domestic and non-domestic buildings like hotels, leisure and sporting facilities, student accommodations, hospitals, etc., it is now common to find showers over baths or shower en-closures as well as wetroom installations with level ac-

Dimension	Values (mm)	Remarks
Diameter of waste outlet	≤ 49	Kitchen sink waste outlet hole diameter 52mm
	≤ 59	Kitchen sink outlet hole – diameter 60mm for stainless steel sinks – diameter 62mm for other sinks
	≤ 87	Kitchen sink waste outlet hole diameter
External diameter of flange	90mm 70 0/-1 52mm/60mm	Kitchen sink waste outlet hole diameter
	85 0/-5	Kitchen sink, waste outlet hole diameter 60mm/62mm
	115 0/-5	Kitchen sink waste outlet hole diameter 90mm
Height of cylindrical part of flange	≤ 1 a)	
Cone angle of contact of waste outlet	≥ 120° a)	
Clamping height of waste outlet	1–6	Kitchen sink – small clamping height
	2–26	Kitchen sink – large clamping height
	44–66	Kitchen sink made of other materials with integral overflow
Outlet connection thread	ISO 228-1-G 1 1/2 B	
	ISO 228-1-G 2 B	
Connection distances	adjustable up to 280	Two-bowl sink
	adjustable up to 400	
Useful length of thread of waste outlet	≥ 11 b)	
Contact diameter of clamping flange	≥ 70	Kitchen sink waste outlet hole diameter 60mm
	≥ 85	Kitchen sink waste outlet hole diameter 62mm
	≥ 110	Kitchen sink waste outlet hole 60mm Kitchen sink waste outlet hole diameter 90mm
Horizontal length from axis of waste outlet to axis of overflow	≥ 120	
Vertical length from waste outlet to axis of overflow	110–180	
Thickness of over-flow	≤ 35	
External dimensions of rectangular overflow	30 0-0,2 58 0-0,2	
External diameter of round overflow	36 0-0,2	
Diameter of overflow barrel	≤ 30	If provided
External diameter of overflow grille	36 0-0,2	
External dimensions of rectangular overflow grille	≤ 30 ≤ 58	
Clamping height of overflow	10–25	Kitchen sink made of ceramics
	1–12	
	2–10	Kitchen sink made of other materials
Thread of nuts	ISO 228-1 G1 1/2	
	ISO 228-1G 2	Kitchen sink with waste outlet hole 90mm

a) If the manufacturer supplies dedicated waste fitting with the kitchen sink, then the values of 'Height of cylindrical part of flange' and 'Cone angle of contact of waste outlet' are not compulsory.
b) The first full diameter thread shall start within 2mm from the end of the spigot.

Table 7.3 Useful dimensions of waste outlets and overflows for kitchen sinks. (Adapted from Table 1, BS EN 274-1 (2002).)

Dimension	Values (mm)	Remarks
Diameter of waste outlet	≤ 42	Wash basin, bidet waste outlet hole diameter 46mm
External diameter of flange	63 0/-3	Wash basin, bidet waste outlet hole diameter 46mm
Height of cylindrical part of flange	≤ 1 a)	--
Cone angle of contact of waste outlet	≥ 110° a)	Wash basin and bidet
Clamping height of waste outlet	8–20 ≥ 40	Wash basin, bidet without integral overflow Wash basin, bidet with integral overflow
Outlet connection thread	ISO 228-1-G 1 1/4 B	Wash basin and bidet
Connection distances	adjustable up to 280 adjustable up to 400	Two-bowl sink
Useful length of thread of waste outlet	≥ 11 b)	--
Contact diameter of clamping flange	≥ 60	Wash basin and bidet with waste outlet hole diameter 46mm
Thickness of over-flow	≤ 35	
External dimensions of rectangular over-flow	30 0-0,2 58 0-0,2	
External diameter of round overflow	36 0-0,2	
Diameter of overflow barrel	≤ 30	If provided
External diameter of overflow grille	36 0-0,2	
External dimensions of rectangular overflow grille	≤ 30 ≤ 58	
Clamping height of overflow	10–25	If made of ceramics
	1–12	If made of ceramics
	2–10	If made of other materials
Thread of nuts	ISO 228-1 G1 1/4	Wash basin and bidet
a) If the manufacturer supplies dedicated waste fitting with the kitchen sink, then the values of 'Height of cylindrical part of flange' and 'Cone angle of contact of waste outlet' are not compulsory. b) The first full diameter thread shall start within 2 mm from the end of the spigot.		

Table 7.4 Useful dimensions of waste outlets and overflows for washbasins. (Adapted from Table 1, BS EN 274-1 (2002).)

cess for users. Lifestyle changes such as walking, running or cycling to work also means that most workplaces now provide showering facilities. In addition, there is a general expectation that adequate spatial and functional provisions are made through good design.

The following standards provide guidelines for the design, specification and installation of showers:

- **BS EN 14527:2016+A1:2018** Shower trays for domestic purposes
- **BS 6340-1:1983** Shower units. Guide on choice of shower units and their components for use in private dwellings
- **BS 6340-2:1983** Shower units. Specification for the installation of shower units
- **BS 6340-4:1984** Shower units. Specification for shower heads and related equipment
- **BS EN 251:2012** Shower trays. Connecting dimensions
- **BS 6465-3** Sanitary installations. Code of practice for the selection, installation and maintenance of san-

Fig. 7.9a.

Fig. 7.9b.

Fig. 7.9c.

Figs 7.9a–c (a) waste outlet with overflow (black convoluted pipe), trap with outlet for cleaning, (b) overflow pipe running downward in front of the hot/cold supply, (c) wastewater pipe with minimal bends connecting downwards to the join with the stack.

Fig. 7.10 Typical domestic drainage set-up showing WC waste to the left and the bath waste with P-trap to the right, the main stack is behind the plasterboard to the right of the WC connector.

Dimension	Values (mm)	Remarks
External diameter of flange	70 0/-1	Bath with waste outlet hole diameter 52mm/60mm
Height of cylindrical part of flange	≤ 1 a)	--
Cone angle of contact of waste outlet	≥ 120° a)	
Clamping height of waste outlet	6–16	Bath with waste outlet hole 52mm
Outlet connection thread	ISO 228-1-G 1 1/2 B	
Useful length of thread of waste outlet	≥ 11 b)	--
Contact diameter of clamping flange	≥ 65	Bath with waste outlet hole diameter 52mm
Horizontal length from axis of waste outlet to axis of overflow	110–170	Special baths
Horizontal length from axis of waste outlet to axis of overflow	170–230	Standard baths
Horizontal length from axis of waste outlet to axis of overflow	> 230	Bath with central waste outlet hole
Vertical length from waste outlet to axis of overflow	330–390	Standard baths
Vertical length from waste outlet to axis of overflow	230–330	Low baths
Vertical length from waste outlet to axis of overflow	390–520	High baths
Thickness of over- flow	≤ 60	
External diameter of round overflow	65 0-0,2	
Diameter of overflow barrel	≤ 49	If provided
External diameter of overflow grille	65–80	
Clamping height of overflow	2–10	
Thread of nuts	ISO 228-1 G1 1/2	
a) If the manufacturer supplies dedicated waste fitting with the kitchen sink, then the values of 'Height of cylindrical part of flange' and 'Cone angle of contact of waste outlet' are not compulsory. b) The first full diameter thread shall start within 2 mm from the end of the spigot.		

Table 7.5 Useful dimensions of waste outlets and overflows for baths. (Adapted from Table 1, BS EN 274-1 (2002) and BS EN 232 (2012).)

itary and associated appliances.

Drainage from a shower could be via a shower tray – raised or flush with the floor or via wet rooms. The trap may also be made out of metallic materials but is typically made of impact-modified extruded acrylic sheet in compliance with BS EN 13558 (2003) Specifications for impact-modified extruded acrylic sheets for shower trays for domestic purposes.

The general guidelines for waste outlets, backflow and overflow still apply. Although a shower tray may be supplied with or without an overflow management option, all should have at least one waste outlet. Water entering the shower outlet should do so in a controlled manner, with the specification of an appropriate screen or enclosure. All water entering the shower tray, with the exception of those retained by surface tension, should drain efficiently via this waste outlet. The shower tray should be mounted in such a manner that ensures the stability of its base, and minimal deflection, especially on loading, prevents poor drainage and structural distortion. Shower trays and floors that reduce the risk of slipping and falling should be specified.

The dimensions of the waste outlet should comply with relevant standards as summarized in Table 7.6. The shower tray should have an appropriate grille or basket strainer to filter objects such as hair, which may enter the drain. It is also beneficial if the trap provides easy access for cleaning out hair and other objects that may cause blockage to the wastewater flow, e.g. by removing the top filter section (Fig. 7.11).

Where wet rooms are installed for level access showers, shower gullies may be preferable to shower traps. Such gullies are specified in European standard BS EN 1253-1:2015 Gullies for buildings. Trapped floor gullies with a depth water seal of at least 50mm. It should be noted that this series of standards is soon to be expanded to include traps of less than 50mm as well as mechanical traps and other variations.

WCs, Toilets

WCs are used to remove liquid and solid human waste from the building. This type of waste is referred to as foul or black wastewater.

The minimum dimensions for WC outlets and other installation dimensions will depend on the type of product (see BS EN 33: 2011 for a wide range of examples and recommended dimensions). Therefore, it is important to consult manufacturer guidelines prior to specification and installation. It is, however, worth noting that no drainpipe serving a WC should be less than 70mm. Some building regulations will permit 70mm pipes from syphonic WCs, however, drainage pipes for WCs are typically 100mm minimum. It is im-

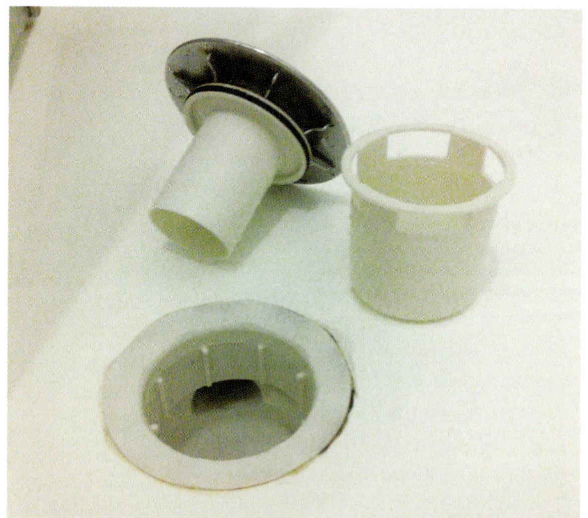

Figs 7.11a–d Sample shower traps for shower trays and wet floors, each with easy access for cleaning.

Dimension	Values (mm)	Remarks
Diameter of waste outlet	≤ 49	Shower tray waste outlet hole diameter 52mm
	≤ 87	Shower tray waste outlet hole diameter 90mm
External diameter of flange	70 0/-1	Shower tray waste outlet hole diameter 52mm/60mm
	85 0/-5	Shower tray waste outlet hole diameter 60mm/62mm
	115 0/-5	Shower tray waste outlet hole diameter 90mm
Height of cylindrical part of flange	≤ 1 a)	
Cone angle of contact of waste outlet	≥ 120° a)	
Clamping height of waste outlet	6–16 6–25	Shower tray waste outlet hole 52mm Shower tray made of ceramics, waste outlet hole 62mm or 90mm
Outlet connection thread	ISO 228-1-G 1 1/2 B	
Contact diameter of clamping flange	≥ 65	Shower tray waste outlet hole diameter 52mm
	≥ 85	Shower tray waste outlet hole diameter 62mm
	≥ 110	Shower tray waste outlet hole diameter 90mm
Horizontal length from axis of waste outlet to axis of overflow	110–170	
Vertical length from waste outlet to axis of overflow	165–260	
Thickness of over-flow	≤ 60	
External diameter of round overflow	65 0-0,2	
Diameter of overflow barrel	≤ 49	If provided
External diameter of overflow grille	65–80	
Clamping height of overflow	2–10	
Thread of nuts	ISO 228-1 G1 1/2	
	ISO 228-1G 2	Shower tray with waste outlet hole 90mm

a) If the manufacturer supplies dedicated waste fitting with the kitchen sink, then the values of 'Height of cylindrical part of flange' and 'Cone angle of contact of waste outlet' are not compulsory.

b) The first full diameter thread shall start within 2mm of the end of the spigot.

Table 7.6 Useful dimensions of waste outlets and overflows for showers. (Adapted from Table 1, BS EN 274-1 (2002) and BS EN 232 (2012).)

portant not to reduce the capacity of a drainage system in the direction of flow, to minimize the risk of blockages. The larger the pipe, the lower the risk of blockage.

In the past WCs have been seen as status symbols (with royalty and the rich) and were often decorated. The flush volumes used many gallons (more than 20l per flush) and WCs were named 'Niagara' and 'The Thunderer' to emphasize their cascades of flush water. Although WCs are sometimes supplied with rainwater or treated wastewater, they still use water that is simply flushed away.

Today, water-using WCs are seen as wasteful of drinking water and so new designs are being produced that treat any human waste substance as a resource, not a waste product. Such 'reinvented toilets', or RTs, are currently only at prototype stages and cost many tens of thousands of pounds, but in time they will become more affordable and could replace the traditional WC in many homes around the world. A switch to such RTs could encourage other appliances to also be viewed as resource-generating items and wastewater systems would need to be totally rethought or replaced.

The WC outlet should be located as close as possible to the foul water discharge stack. The discharge stack should be ventilated using one of the systems/strategies discussed earlier.

The following standards provide guidelines for the design, specification and installation of WCs:

- **BS EN 997:2018** WC pans and WC suites with integral trap
- **BS EN 14055:2018** WC and urinal flushing cisterns
- **BS EN 33:2011** Connecting dimensions for WC pans and WC suites
- **BS EN 14055:2018** WC and urinal flushing cisterns
- **BS 6465-3** Sanitary installations. Code of practice for the selection, installation and maintenance of sanitary and associated appliances.

Figs 7.12a-b The Propelair WC uses air in addition to water to flush the pan. To ensure that all the force of the flush displaces the waste matter in the bowl, the WC can only be flushed once the lid is lowered and locked closed. This WC only uses 1.5 litres of water per flush. It connects to a conventional drainage system. The WC can be supplied with an opaque or transparent lid, depending upon the customer's requirements.

Figs 7.13a–b (a) Pedestal WCs showing visible and (b) hidden cisterns and waste outlets.

WC pans are typically made out of glazed ceramic or vitreous china but the WC cistern may be made out of a variety of materials. There are different types of WCs with variations of: the cistern design, mounting mechanism, flushing mechanism, backflow and overflow control mechanisms. WCs can be wall hung or pedestal mounted. The cistern and bowl/pan can be an integrated unit or separated. The units can be entirely visible or the cistern can be fully or partially concealed. The cistern is used to deliver the right amount of water required for flushing. The cistern will have an inlet and outlet that is controlled by a pressure or mechanical valve. Modern cisterns now permit at least two measured amounts of water to be delivered to the bowl for efficient water use during flushing, i.e. a full flush or reduced flush. All cisterns should have an overflow management mechanism. Before 1999, in the UK, most WC cisterns had an external overflow pipe that would direct any excess water to outside of the building where it could be visible to the owner and the water company's inspector. Since then, internal overflows have been permitted. These direct any excess water through the cistern's outlet and into the WC pan, where it can be seen by the users. Although the water from a leaking water inlet valve is not an issue for the drainage system, it is a waste of water and should be rectified as soon as possible. Although all WCs originally had flap or drop outlet valves, for many years siphons were used in preference as they were also known as 'wastewater preventers', or WPPs, as they could not leak. However, due to the volume of water needed to prime them and make them operate they were not readily adaptable to low-volume flush and flap and drop valves were reintroduced. The new outlet valves were far superior to the original valves (they were made of leather) and have to pass extensive endurance tests to meet the CEN standards. Even though they are good, they have gained a reputation for leaking, hence the market has now introduced advanced

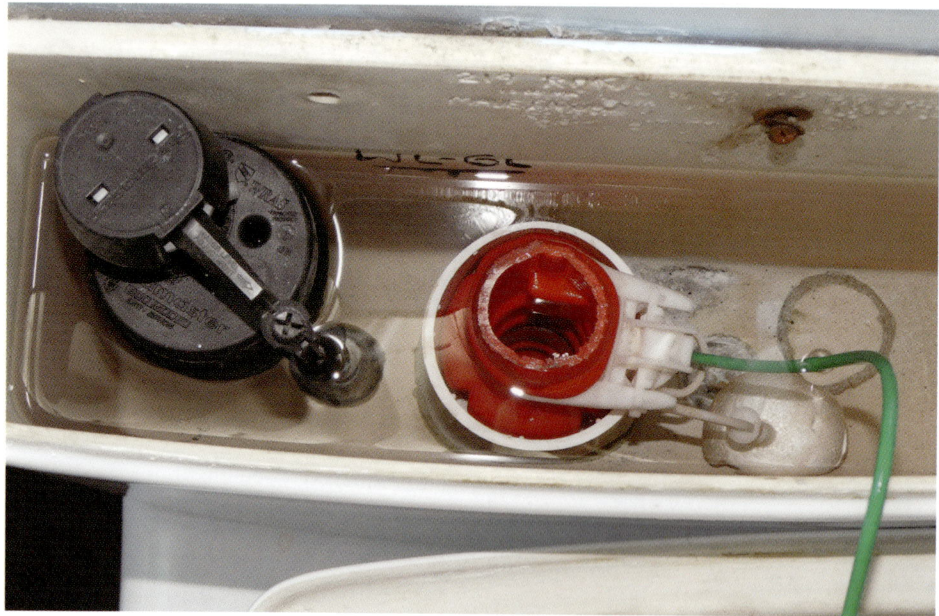

Fig. 7.14 (a) The WC cistern with a syphon is a typical low-level plastic cistern with a traditional water inlet valve float-operated valve. This cistern has an overflow pipe to the side that can be taken to an external overflow. (b) The WC cistern with the red outlet valve is in a typical close-coupled ceramic cistern with the operating buttons located in the top of the lid. The dual-flush buttons have different lengths so that for a short flush the actuating lever is deflected through a smaller arc than when the long-flush button is pressed. The actuating lever is connected to the outlet valve using a Bowden-type cable. The top of the outlet valve is open to create an internal overflow. A compact equilibrium water inlet valve is fitted.

WC cistern syphons that will work with low volumes and should waste less water. For commercial installations, WC flushing valves are available that run directly from a mains supply and need no cisterns. The size of water main required excludes them from domestic installations.

The WC pan typically includes an integral water trap to prevent the backflow of foul air from the pipes. This is why most WC bowls should retain some amount of water after flushing. The amount of water retained will vary according to the design of the product, i.e. how the system deals with the waste that falls into the pan:

Washdown – waste falls directly into the sump of the bowl and is then removed by the flushing water

Washout – waste falls onto a small shelf towards the back of the pan and is then flushed off the shelf into the sump of the bowl and finally removed by the flushing water.

Siphonic (single trap) – the waste falls directly into the bowl, but when the flush commences a proportion of the flush is diverted to a venturi device that creates a reduced pressure zone in the crown of the trap and hence the waste is then siphoned out.

Siphonic (double trap) – the waste falls directly into the bowl, but the pan contains two shallow traps in sequence, so that as the water flows through the pan there is a low resistance but the first trap creates a restriction so that the water in the bowl builds up until there is sufficient head for a syphonic action to take place.

Siphonic (low volume flush) – the previous types of siphonic WC are quiet in operation, but tend to use high volumes of water. The double trap is also particularly prone to blockage. This type of WC pan resembles a conventional washdown but with a resilient constriction on the outlet. When the flush commences, the water is backed up by the outlet until there is sufficient head for the constriction to dilate and pass the flush rapidly. After the flush, the constriction resumes its partially closed position.

However, the depth of the (total) water seal should be no less than 50mm. If no water is retained after any of these actions, the trap seal is broken and should be investigated.

Combination WCs (with wash and dry)

Although bidets were common in the late twentieth century, due to space and plumbing requirements they are now a rare item in modern bathrooms. One item that is tending to replace them is the WC with built-in wash and dry facilities. These WCs require not only a water supply and drainage, but also electricity. Early combination WCs were only permitted in exceptional circumstances, due to their inherent back-siphonage risks. The water supplies had to be the same for an ascending spray bidet with a break cistern at a suitable height to feed the wash jets. Newer combination WCs have retractable spray arms that are above the flood level of the pan and can be programmed to clean and massage. The typical price for a combination WC is around £5,000–£7,000, making them rather exclusive.

Washing Machines and Dishwashers

Washing machines are used for cleaning clothes and other laundry materials, while dishwashers are useful for cleaning cooking and eating utensils and crockery.

Wastewater from washing machines and dishwashers typically discharge via a flexible hose, typically 22mm in diameter, through a trap to the wastewater pipe and then the ventilated discharge stack. Appliances and fittings are always connected before and not after the trap. If the machine is being installed next to a kitchen sink, then the hose connects directly via a screw compression spigot and connector between the sink waste and the trap. In this instance, the top of the discharge hose should be near the top level of the unit to prevent backflow or wastewater stagnation (and

Fig. 7.15 Washing machine/dishwasher connection under a sink with a P trap.

subsequent odours) in the pipe. Multiple bends in the hose should be avoided for this reason.

If not discharged via the sink assembly, a vertical pipe must be installed that then discharges via a trap to the soil stack. The hose is then connected loosely to a vertical waste pipe or a separate ventilating stack provided that the connection between the flexible hose and the vertical pipe is sealed. Although older machines tended to simply fill, wash and discharge, modern machines use a large number of small quantities of water and discharge them in short bursts. So the loading on the drainage system from washing machines, and similarly dishwashers, is less than it used to be despite the capacity of machines increasing.

Wastewater from washing machines and dishwashers should be discharged to the wastewater, and not the rain or surface water drainage system.

Each sanitary fitting and appliance generates waste or foul water when used that needs to be channelled and discharged safely from the building into the sewer system. Although they serve different functions, their discharge needs are summarized as:

1. The effective collection and channelling of the wastewater from the appliance into the relevant pipework, stack and onwards to the sewer or for recycling and reuse.

2. The prevention of backflow of contaminated water and foul air from the appliance or pipework into the potable water network and/or the building.

3. The prevention of overflow of the appliance or fitting such that water or wastewater overflows do not damage the building.

4. The prevention of solids and other undesired contaminants entering into the internal drainage system.

5. The provision of access for cleaning or maintenance of the outlets, connections, traps and pipework.

These factors should be considered and applied to assure the good functioning of any internal drainage system.

Chapter Summary

The start of any drainage system is the appliance that it drains. Sanitary appliances by their use are the main items that produce wastewater. There are normally overflow systems and water-sealed traps within each appliance. The water normally drains from the appliance via a 'waste fitting' that connects the appliance to the trap. Such waste fittings can vary greatly in design and operation.

The design, shape and volume of the appliance can influence the amount of wastewater that is discharged and how it is discharged. For example, a washbasin with an unplugged outlet and a spray tap will discharge at the same rate as the tap flow for the duration of the operation of the tap. A basin with a plug fitted will not discharge until the plug is removed and then at a rate greater that the flow from the tap, but only normally a volume up to the overflow of the appliance. Similarly, a bath with a shower may operate as a shower tray, with the discharge linked to the flow from the showerhead, or as a bath, with a large volume discharge after the bath has been used.

The appliance with the greatest impact on a drainage system is normally the WC. However, over the last three or four decades the volume flushed, at each operation, has been considerably reduced. If this trend continues, the design of drainage systems may be significantly changed.

Chapter 8

INTEGRATION AND INNOVATION

Introduction

Drainage systems are designed to manage the collection, discharge, removal and/or treatment, recycle or reuse of waste, foul and surface water in and around buildings. This book covers, to a useful level of detail, the important design considerations for building drainage with particular emphasis on internal plumbing and on-site drainage systems. It focuses on domestic buildings as a template for explaining the key principles and techniques available to architects, architectural technologists, other building designers and professionals and the general public.

It emphasizes the important design considerations to achieve the effective functioning and performance of drainage systems, minimize failures and problems, achieve efficient discharge and flow, maintain good health and comfort for building users and deliver durable, sustainable and resilient drainage systems. The preceding chapters present the important aspects of building drainage as separate sub-sets of the whole. This chapter concludes the book by highlighting the importance of an integrated approach to achieve innovative drainage design and systems.

Integration

Integrated design is a collaborative design approach aimed at delivering a high-value, high-performance, sustainable and resilient building through the integration and optimization of every aspect, including its systems. To be successful, integration must commence at the early design stages, through to construction, operation, occupancy and finally decommissioning.

Future-proofing should be considered alongside the life cycle design issues. The building or development needs to be designed in such a way that it can easily be adapted and drainage systems enhanced as new innovations emerge or as the purpose and functions of the buildings evolve.

Integration for drainage systems should consider the following six principles (Fig. 8.1):

1 **Site, building, environment:** The site is an important consideration in drainage design. Climate and weather, soil type and condition, rural or urban, proximity to existing drainage networks, etc., all inform the approach to drainage design, as well as the type and specification of the drainage solution. The size of the site, nearby buildings and functions, landscaping needs, planning and regulatory standards for different areas will also inform what is feasible or practicable for a specific site or building.

Fig. 8.1 Integrated design factors for drainage.

The building type also informs the type and scale of drainage design. The solutions for a bungalow will differ from that of a seven-storey apartment block, office building or factory.

2   **Resources:** Note that the term 'resource' is used here, rather than waste. It is important that rain, surface and wastewater are considered as resources of value from the early design stages. Drainage system design should therefore consider on-site circularity of these resources rather than designing for 'capture and discharge' only. Rainwater, with minimal treatment, can be reused for non-potable uses on site, e.g. garden irrigation, toilet flushing or even managed reinfiltration into the ground to recharge underground aquifers. The latter will help to reduce flooding events and to minimize soil erosion, degradation and droughts. Energy can be recaptured directly from wastewater discharging from showers, etc., and reused for space heating or pre-

heating a supply to a water heater. Wastewater can be captured and treated as a means for delivering added building value, delivering sustainable water with less carbon footprint, affordable non-potable water for communal use and so on. There are many schemes in Europe that have done this successfully. However, this happened through integration and collaboration from project inception, not as a bolt-on at the later stages. Ultimately, the ambition should be net-zero discharge, similarly to net zero-carbon or energy buildings.

3   **Design factors:** It is important to have good architectural design for drainage; likewise, poor drainage will impact on good architectural design. Adequate space is required for drainage in the design. This includes access for maintenance, repairs and adaptations. Plant and storage are often overlooked in architectural design, with spaces and provisions typically undersized in most buildings due to other spaces being given more priority or considered more important. However, adequate storage and plant space is required for effective (and to avoid aesthetically unsightly) wastewater management, as well as to facilitate energy and wastewater recovery and reuse when and where appropriate. Prior to drainage design, the following must also be considered: water supply (type, size, scale and amount of); performance and functional specifications of water fittings and appliances, and their location. Lastly, a robust understanding of standards, regulations and laws governing the design of the building drainage system should be considered as this may vary per site, region, country, etc.

4   **Process:** The design process should be as collaborative as is feasible for the scale and type of project. It is beneficial for architectural professionals to consult other relevant professionals, including civil and structural engineers, M&E engineers, geologists and landscape architects, and to make adjustments

as appropriate. This book helps to understand some of the rationale behind the design advice and recommendations that may be received during this collaborative design process. It may be necessary to adjust spatial configurations to afford straight drainage runs, make room for stacks, or storage tanks, etc., to ensure adequate structural support, and avoid conflicts with structural elements, to minimize the impact of drainage on site access, landscaping and aesthetics; or to integrate and optimize water and drainage systems to deliver a more sustainable and environmentally resilient building, e.g. to recycle water as well as minimize flooding. The collaborative design process should also ensure that decisions do not originate and stop at design. Instead, the specification processes should account for installation needs, opportunities and constraints, especially during the building construction phase. This should be followed by management, maintenance, adaptation, decommissioning and upgrade during the building in-use phase.

5 **Technology** – Materials, products, systems: The principles of building drainage, e.g. the need for unobstructed flow, are simple and straightforward, and have not changed much for centuries. However, the materials, products and systems with which these principles are implemented have evolved, and will continue to evolve and improve. It was common practice to use clay, concrete and even lead in wastewater and rainwater systems. Nowadays, different types of plastic, and some metals, are more commonly used. Similarly, combined sewer systems were common. However, with more knowledge and increasing need to better utilize and manage surface water flows, the new trend, especially in the UK, is towards a separate sewer system. Other innovations in drainage systems have also been discussed. For instance, the use of pumped or vacuum systems may be necessary where the baseline gravity systems may be inefficient or unviable. There are also a number of emerging smart technologies that may be embedded with drainage systems, e.g. for real-time monitoring of system use and performance, flagging maintenance and repair needs. Therefore, it is imperative for all architectural professionals to keep up to date, not just with design trends, standards and regulations, but also with new and emerging technologies.

6 **People:** Poor drainage design can impact on the health, comfort and well-being of the building users. Microbial, foul air and noise encroachment into buildings are common complaints by building users and this can easily be avoided through good architectural and drainage design. The health risks and susceptibility of users should be considered in drainage design. The approach to drainage for an office building will be different to that for a hospital or care home. The number and type of users, as well as their known or anticipated needs, behaviours and requirements, will inform the drainage design. For instance, the occupancy number and pattern of use will help to make informed decisions about whether to invest in rain or wastewater recovery and reuse systems or heat recovery systems. The latter makes a lot of sense in a hotel or leisure centre, but may not be cost-effective in a single occupancy-dwelling, especially if being retrofitted.

Innovation and Future Systems

Integrated drainage systems are now being introduced and developed all around the world. Innovation and future systems are typically discussed in technological terms. However, true integrated solutions combine the six principles previously discussed. Nonetheless, smart systems are worth a mention here.

Smart Systems

Efficient drainage design starts with understanding the quantities and qualities of substances to be managed. Information is therefore key; this makes it useful to measure what goes in or out of a specified site boundary. Smart technologies including tagging, sensors, data transmission and logging devices, as well as information analytics software, are now increasingly installed in buildings as part of a separate or integrated building information and management system. Smart devices associated with drainage systems include:

- Smart water meters
- Smart energy meters
- Smart heating and cooling
- Smart WCs
- Smart wastewater controls
- Smart rainwater management/harvesting systems
- Smart flood controls and management systems.

It is not easy, but also not impossible to integrate the various smart systems. To be effective, the systems need to be intuitive, configurable and adaptable, as they often require frequent updates, improvements and replacement. These sensors can also become obsolete as the technologies and protocols advance. As new generations of device come to market – with greater attributes, abilities and flexibility; along with reductions in size and cost – the old ones become another waste stream. So, future-proofing should be considered with strategies put in place. The strategies should include decommissioning protocols such as how to recycle the drainage materials with its embedded electronics.

Advantages of smart integrated water systems include:

- Ultra-efficient water use
- Shallow, easy-access, drainage pipe networks
- Drains that do not rely upon water for transportation, but use air, or an even better transport fluid
- No reliance upon gravity, so transportation can be in any direction
- Separate pipes for different waste/resource streams
- Sustainable building-based water cycle systems: where all waste/resources are dealt with at source and the only emissions and products to leave the building are:
  - Usable energy, drinkable water, non-drinkable water, surplus surface water and separated waste/resource streams

The Future

The key element to future systems is a multi-pass system, not the common single-pass traditional system. Recycling wastewater is not new, but the proportion will need to rise considerably in the future. To achieve this, the accepted technology will need to advance. The current obstacles to the adoption and acceptance of water reuse systems include:

- Visual appearance of water
- Risk of cross-connections and contamination of wholesome water supply
- Volume needed for reuse equipment
- Consumables and maintenance regimes
- Perceptions of unsafe or unpleasant water
- Wide range of performances
- Expensive when compared to cost of mains water
- Lack of confidence in systems from most plumbers/installers
- Impact on building's value
- Inconsistent world standards for reuse pipework.

Meanwhile, since around the end of the last century the development of waterless traps has shown that flexible tubing can be used to transport waste. Many of these new mechanical traps are modelled upon human organs, such as the eyelids and digestive track components. As with SuDS, where biomimicry has predominated, there is potential to design a wastewater system along the lines of a human, or animal, digestive system.

By using flexible pipes that can expand and contract, by a number of factors, loading of networks can become far more variable without loss of performance. Containing fluids within energy fields, instead of pipes, is an intriguing option. Currently, the energy required to do so would be prohibitive and the consequences of a power failure could be very messy. If energy is not a problem, vaporization of liquid waste could be used and the resulting gases transported by gas pipes or gas containers; if transport is actually needed. This can be implemented at different scales:

- **Small scale reuse** – as in Canada's Toronto House – on an individual dwellings scale
- **Medium scale reuse** – as in Japan's Tokyo skyscrapers with full waste and water recycling facilities in their basements – on a whole building or community basis
- **National scale** – as in Israel's irrigation system – creating a nationwide infrastructure to reuse wastewater
- Or, a combination of scales.

In Conclusion

Although drainage has changed little in hundreds of years, there are now opportunities and drivers to change drainage design. The integration of water supplies, stormwater management and wastewater treatment should bring far more sustainable systems where there is no longer any 'waste' but just different 'resources'.

We have seen significant changes over the last few decades with the development of SuDS, with green and blue engineering approaches, so we can now expect similar policies and practices to be extended to the whole water cycle. As appliances become more efficient and generate less and less water to be disposed of, the whole concept of taking away water from a building will be challenged. Although such schemes may be incorporated into new cities and buildings, there will still be the vast majority of buildings in developed countries that have the drainage systems they were designed with, so maintaining and avoiding a negative impact on existing systems will be required for some time.

INDEX